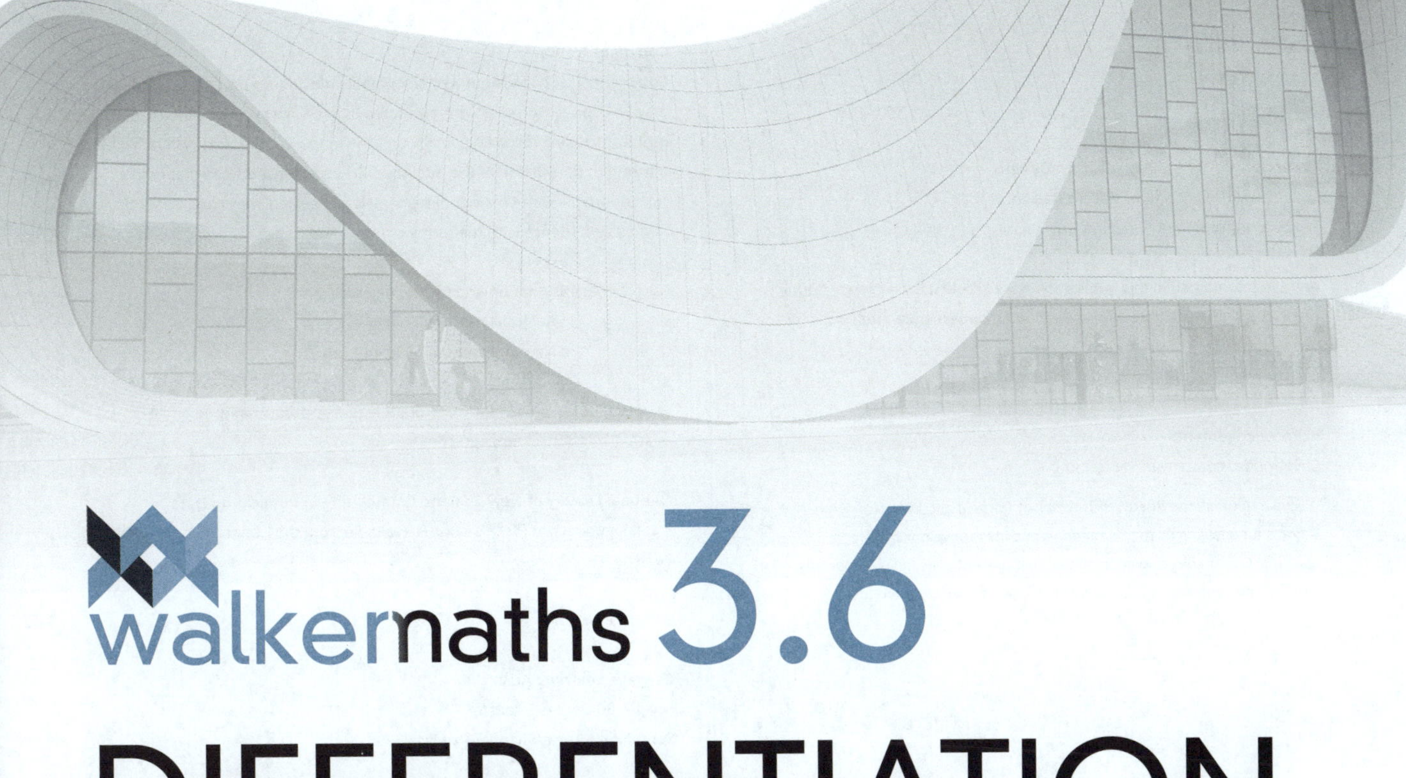

walkermaths 3.6

DIFFERENTIATION

NCEA Level 3 Internal

Charlotte Walker and Victoria Walker

Walker Maths 3.6 Differientiation
1st Edition
Charlotte Walker
Victoria Walker

Cover design: Cheryl Smith, Macarn Design
Text design: Cheryl Smith, Macarn Design
Production controller: Siew Han Ong

Acknowledgements
Cover photo courtesy of Shutterstock

We wish to thank the Boards of Trustees of Darfield and Riccarton High Schools for allowing us to use materials and ideas developed while teaching. Our thanks also go to all past and present colleagues, especially Kath Wilson, who have generously shared their experience and ideas.

For product information and technology assistance,
in Australia call **1300 790 853**;
in New Zealand call **0800 449 725**

For permission to use material from this text or product, please email **aust.permissions@cengage.com**

National Library of New Zealand Cataloguing-in-Publication Data
A catalogue record for this book is available from the National Library of New Zealand

9780170446976

Cengage Learning Australia
Level 7, 80 Dorcas Street
South Melbourne, Victoria Australia 3205

Cengage Learning New Zealand
Unit 4B Rosedale Office Park
331 Rosedale Road, Albany, North Shore 0632, NZ

For learning solutions, visit **cengage.co.nz**

Printed in China by 1010 Printing International Limited.
5 6 7 26

CONTENTS

ISBN: 9780170446976

Formulae

These are the relevant formulae that are supplied to you in the external examination:

Algebra

Logarithms

$y = \log_b x \Leftrightarrow x = b^y$

$\log_b(xy) = \log_b x + \log_b y$

$\log_b\left(\frac{x}{y}\right) = \log_b x - \log_b y$

$\log_b(x^n) = n\log_b x$

$\log_b x = \frac{\log_a x}{\log_a b}$

Coordinate geometry

Straight line

Equation $y - y_1 = m(x - x_1)$

Calculus

Product rule

$(f.g)' = f.g' + g.f'$ or if $y = uv$

then $\frac{dy}{dx} = u\frac{dv}{dx} + v\frac{du}{dx}$

Quotient rule

$\left(\frac{f}{g}\right)' = \frac{g.f' - f.g'}{g^2}$ or if $y = \frac{u}{v}$

then $\frac{dy}{dx} = \frac{v\frac{du}{dx} - u\frac{dv}{dx}}{v^2}$

Composite function or chain rule

$(f(g))' = f'(g).g'$

or if $y = f(u)$ and $u = g(x)$ then $\frac{dy}{dx} = \frac{dy}{du} \cdot \frac{du}{dx}$

Parametric function

$\frac{dy}{dx} = \frac{dy}{dt} \cdot \frac{dt}{dx}$

$\frac{d^2y}{dx^2} = \frac{d}{dt}\left(\frac{dy}{dx}\right) \cdot \frac{dt}{dx}$

Differentiation

$y = f(x)$	$\frac{dy}{dx} = f'(x)$
$\ln x$	$\frac{1}{x}$
e^{ax}	ae^{ax}
$\sin x$	$\cos x$
$\cos x$	$-\sin x$
$\tan x$	$\sec^2 x$
$\text{cosec}\, x$	$-\text{cosec}\, x \cot x$
$\sec x$	$\sec x \tan x$
$\cot x$	$-\text{cosec}^2 x$

Measurement

Triangle

Area $= \frac{1}{2}ab\sin C$

Trapezium

Area $= \frac{1}{2}(a + b)h$

Sector

Area $= \frac{1}{2}r^2\theta$

Arc length $= r\theta$

Cylinder

Volume $= \pi r^2 h$

Curved surface area $= 2\pi rh$

Cone

Volume $= \frac{1}{3}\pi r^2 h$

Curved surface area $= \pi rl$

where l = slant height

Sphere

Volume $= \frac{4}{3}\pi r^3$

Surface area $= 4\pi r^2$

ISBN: 9780170446976

Glossary

Make your own glossary of key terms:

Term	Definition	Picture/Example
Coefficient		
Exponent		
Constant		
Polynomial		
Domain		
Range		
Differentiation		
Anti-differentiation		
Integration		
Derivative		

Term	Definition	Picture/Example
Asymptote		
Inverse		
e		
Log		
Parameters		
Optimisation		
Maximum (plural: maxima)		
Minimum (plural: minima)		
Kinematics		
Velocity		
Acceleration		

ISBN: 9780170446976

Revision

Terminology

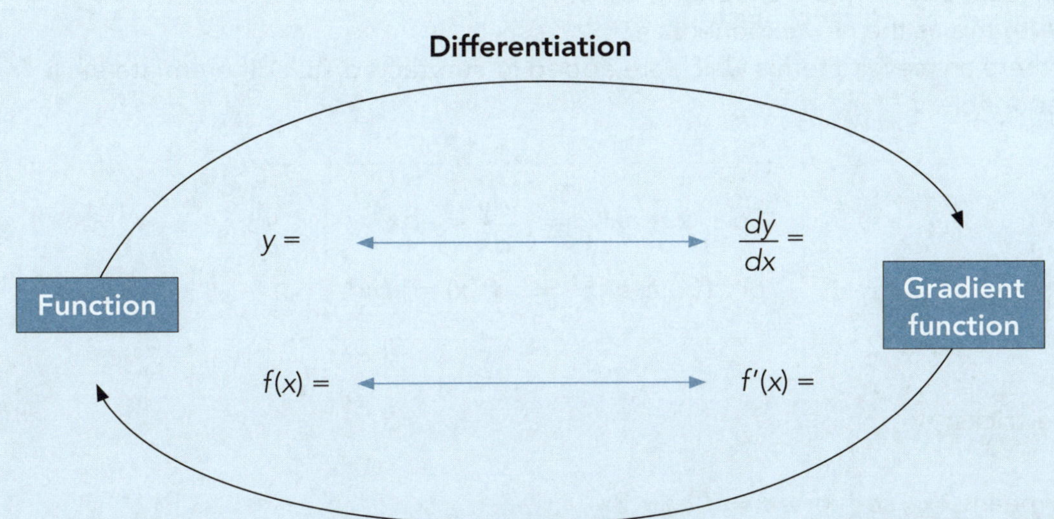

$\frac{dy}{dx}$ is said 'd y by d x' and means 'the derivative of y with respect to x' or $\frac{d}{dx}$ of y.

This is known as 'Leibniz' notation.

Conventions

If the function is expressed as y = ..., then:

the gradient function is expressed as $\frac{dy}{dx} = \ldots$, e.g. $y = 2x^3 \Rightarrow \frac{dy}{dx} = 6x^2$

and the second derivative is expressed as $\frac{d^2y}{dx^2} = \ldots$, e.g. $y = 2x^3 \Rightarrow \frac{d^2y}{dx^2} = 12x$

$\frac{d^2y}{dx^2}$ is said 'd 2 y by d x squared' and means 'the second derivative of y with respect to x' or $\frac{d}{dx}$ of $\frac{dy}{dx}$.

This is known as 'function' notation.

If the function is expressed as f(x) = ..., then:

the gradient function is expressed as $f'(x) = \ldots$, e.g. $f(x) = 2x^3 \Rightarrow f'(x) = 6x^2$

and the second derivative is expressed as $f''(x) = \ldots$, e.g. $f(x) = 2x^3 \Rightarrow f''(x) = 12x$

ISBN: 9780170446976

Differentiation of polynomials

The rule: Let $y = ax^b$.

1 Multiply the coefficient by the exponent: $a \times b = ab$.
Write this as the new coefficient.
2 Subtract one from the exponent: $b - 1$.
Write this as the new exponent.
3 If there are several terms which are added or subtracted, just differentiate each separately.

$$\text{So} \quad \mathbf{y = ax^b \;\Rightarrow\; \frac{dy}{dx} = abx^{b-1}}$$

$$\text{or} \quad \mathbf{f(x) = ax^b \;\Rightarrow\; f'(x) = abx^{b-1}}$$

Some tricks:

1 x^3 means $1x^3$, so $f(x) = x^3 \Rightarrow f'(x) = 3x^2$

2 x means x^1 and $x^0 = 1$, so $f(x) = 7x \Rightarrow f'(x) = 7$

Provided $x \neq 0$

3 Constants (e.g. 4) can be written as $4x^0$, so $f(x) = 4 \Rightarrow f'(x) = 0$

4 Fractions: $\frac{x^2}{3}$ can be written as $\frac{1}{3}x^2$, so $f(x) = \frac{x^2}{3} \Rightarrow f'(x) = \frac{2}{3}x$

5 Brackets: expand these before differentiating:

$$\text{so } f(x) = \frac{1}{4}(20 - x) = 5 - \frac{1}{4}x \Rightarrow f'(x) = -\frac{1}{4}$$

Examples:

1 $$y = 5x^4 - \frac{2x^3}{3} - x \Rightarrow \frac{dy}{dx} = 20x^3 - 2x^2 - 1$$

2 $$f(x) = (3x + 4)(2 - x^2) = 6x - 3x^3 + 8 - 4x^2$$
$$= -3x^3 - 4x^2 + 6x + 8$$
$$\Rightarrow f'(x) = -9x^2 - 8x + 6$$

ISBN: 9780170446976

Differentiate the following polynomials.

1 $y = \frac{2x}{3}$

2 $f(x) = \frac{3x^2}{8} + 5x$

3 $y = 7x^2 - \frac{2x^6}{3}$

4 $f(x) = \frac{4x^2 + x}{8}$

5 $y = 5x(7 - 4x)$

6 $f(x) = -3(2x - 1)(5x + 7)$

7 $y = \frac{-x^3 - 7}{8}$

8 $f(x) = 0.01(3x^2 + 8x - 2)$

9 $y = 10 - 0.2(3x - 7)$

10 $f(x) = \frac{5(x^3 - 2x^2)}{2}$

11 $y = 7 - \frac{x^5 - 3x}{5}$

12 $f(x) = x(2x - 7)^2$

Logarithms

Fundamentals

- Logarithms ('logs') and exponents are very closely related:

This is given to you on the formula sheet.

$$b^x = y \Leftrightarrow \log_b y = x$$

In words: 'log to the **base** b of y equals x'.

Example:	**Exponent form**		**Log form**
	$10^2 = 100$	$\Rightarrow$	$\log_{10} 100 = 2$
	$3^4 = 81$	$\Rightarrow$	$\log_3 81 = 4$
	$2^7 = 128$	$\Rightarrow$	$\log_2 128 = 7$
	$5^a = b$	$\Rightarrow$	$\log_5 b = a$

Notice that the power (exponent) to which the base is raised is the same as the value of the log.

Note:	$10^3 = 1000$	$\Rightarrow$	$\log_{10} 1000 = 3$
		or	$\log_{10} 10^3 = 3$

If these match, the value of the log is the same as the power.

Write the following in log form.

1 $2^4 = 16$ ________________

2 $10^6 = a$ ________________

3 $3^b = 9$ ________________

4 $5^1 = 5$ ________________

5 $c^3 = d$ ________________

6 $a^b = c$ ________________

Write the following in exponent form.

7 $\log_2 a = 5$ ________________

8 $\log_b c = 4$ ________________

9 $\log_b c = d$ ________________

10 $\log_d 125 = b$ ________________

11 $\log_x y = z$ ________________

12 $\log_c a = b$ ________________

ISBN: 9780170446976

Some tricks:

		Examples: Exponent form	Examples: Log form
1	$\log_b 1 = 0$ because $(\text{anything})^0 = 1$	$8^0 = 1$	$\log_8 1 = 0$
2	$\log_b b = 1$ because $b^1 = b$	$17^1 = 17$	$\log_{17} 17 = 1$
3	$\log_b \frac{1}{b^x} = -x$ because $\frac{1}{b^x} = b^{-x}$	$\frac{1}{16} = \frac{1}{2^4} = 2^{-4}$	$\log_2 \frac{1}{16} = \log_2 2^{-4} = -4$
4	$\log_b(\sqrt[n]{b^m}) = \frac{m}{n}$ because $(\sqrt[n]{b^m}) = b^{\frac{m}{n}}$	$4 = \sqrt[3]{8^2} = 8^{\frac{2}{3}}$	$\log_8 4 = \log_8 (8^{\frac{2}{3}}) = \frac{2}{3}$

Estimating the value of a log, e.g. $\log_4 50$.

$\left.\begin{array}{l}\log_4 16 = 2 \\ \log_4 64 = 3\end{array}\right\}$ so $\log_4 50$ must lie between 2 and 3.

Write the following in log form.

13 $9^0 = 1$ __________

14 $7^1 = 7$ __________

15 $10^{-4} = 0.0001$ __________

16 $25^{\frac{1}{2}} = 5$ __________

17 $6^{-1} = \frac{1}{6}$ __________

18 $16^{\frac{1}{4}} = 2$ __________

19 $25^{-\frac{1}{2}} = \frac{1}{5}$ __________

20 $81^{0.5} = 9$ __________

Write the following in exponent form.

21 $\log_5 1 = 0$ __________

22 $\log_5 \frac{1}{5} = -1$ __________

23 $\log_{49} 7 = \frac{1}{2}$ __________

24 $\log_{10} 10 = 1$ __________

25 $\log_{64} 2 = \frac{1}{6}$ __________

26 $\log_{10} 0.01 = -2$ __________

27 $\log_{10} \frac{1}{10\,000} = -4$ __________

28 $\log_2 \frac{1}{8} = -3$ __________

29 *Without* using a calculator, match each log statement to its correct value.

Log
$\log_5 200$
$\log_{100} 14$
$\log_{10} 200\,000$
$\log_2 80$
$\log_7 60$
$\log_3 90$
$\log_{23} 30$

Value
0.573
4.096
3.292
5.301
1.085
6.322
2.104

ISBN: 9780170446976

Rules

- These are all on your formula sheet.

1 $\log a + \log b = \log ab$

Examples:

1 $\log 7 + \log 3 = \log 21$

2 $\log 4 + \log 5 + \log 6 = \log 120$

Write the following as the log of a single number or expression.

1 $\log 5 + \log 7$ ________

2 $\log 20 + \log 3$ ________

3 $\log 3 + \log 5 + \log 2$ ________

4 $\log 1 + \log 10$ ________

5 $\log 1 + \log 1 + \log 1$ ________

6 $\log p + \log q + \log r$ ________

Write the following in terms of logs of separate numbers or variables.

7 $\log pq$ ________

8 $\log abc$ ________

2 $\log a - \log b = \log \frac{a}{b}$

Examples:

1 $\log 18 - \log 3 = \log \frac{18}{3}$
$= \log 6$

2 $\log 8 + \log 5 - \log 10 = \log \frac{8 \times 5}{10}$
$= \log 4$

Write the following as the log of a single number or expression.

9 $\log 15 - \log 3$ ________

10 $\log 15 - \log 1$ ________

11 $\log 5 + \log 6 - \log 3$ ________

12 $\log f + \log g - \log h$ ________

Write the following in terms of logs of separate numbers or variables.

13 $\log \frac{24}{b}$ ________

14 $\log \frac{c}{2}$ ________

15 $\log \frac{27}{2 \times d}$ ________

16 $\log \frac{n \times p}{q}$ ________

ISBN: 9780170446976

3 $n\log a = \log a^n$

Examples:

1 $3\log 5 = \log 5^3$

$= \log 125$

2 $\frac{1}{4}\log 16 = \log 16^{\frac{1}{4}}$

$= \log 2$

Write the following as the log of a single number.

17 $2\log 9$ ________________

18 $\frac{1}{2}\log 81$ ________________

19 $\frac{1}{3}\log 27$ ________________

20 $\frac{1}{a}\log b$ ________________

Putting the rules together

- It is often useful to rewrite the expression using the logarithm of the smallest number possible.
- Use the logarithm laws to simplify the expression.
- There can be several ways of doing these.

Examples:

1

$$\begin{aligned}\frac{1}{4}\log 81 - \log 15 + \log 5 &= \log \sqrt[4]{81} - \log 15 + \log 5 \\ &= \log 3 - \log 15 + \log 5 \\ &= \log \frac{3 \times 5}{15} \\ &= \log 1 \\ &= 0\end{aligned}$$

2

$$\begin{aligned}\frac{\log 8 + \frac{1}{3}\log 64}{3\log 4} &= \frac{3\log 2 + \log 4}{3\log 2^2} \\ &= \frac{3\log 2 + 2\log 2}{6\log 2} \\ &= \frac{5\log 2}{6\log 2} \\ &= \frac{5}{6}\end{aligned}$$

Express all parts of the expression in terms of log 2.

ISBN: 9780170446976

Simplify the following.

21 $\dfrac{\log 81}{\log 3}$

22 $\log 64 - \dfrac{1}{2}\log 8$

23 $\dfrac{\frac{1}{2}\log 9}{\log 243}$

24 $\dfrac{1}{3}\log 125 + 2\log 5 - \log 25$

25 $\dfrac{2(3\log 2 + \log 16)}{3\log 4}$

26 $\dfrac{1}{2}\log 81 + \log\left(\dfrac{1}{9}\right)$

27 $\dfrac{\log 128 - 3\log 2^2}{\log 8}$

28 $\dfrac{3\log 10 + \frac{1}{2}\log 100}{\log 10^6}$

29 $\log\left(\dfrac{1}{100}\right) + 3\log 10^{-3}$

30 $\dfrac{1}{2}\log\left(\dfrac{1}{16}\right) + \dfrac{1}{3}\log 64$

ISBN: 9780170446976

Exact values for trigonometric functions: special triangles

- Trigonometric functions for **most angles do not have exact values**.
- Your calculator gives you these values to many decimal places, and it is usual to **round** these to **4 decimal places**.
  For example: $\cos 0.6 = 0.825335614... = 0.8253$ (4 dp)
- Using **special triangles**, it is possible to write **exact values** for some angles.
- **Exact** values must be able to be written as fractions made up of **whole numbers** and/or **surds** (roots of whole numbers).

For $\frac{\pi}{4}$ (or 45°), use an isosceles right-angled triangle with equal sides 1 unit long:

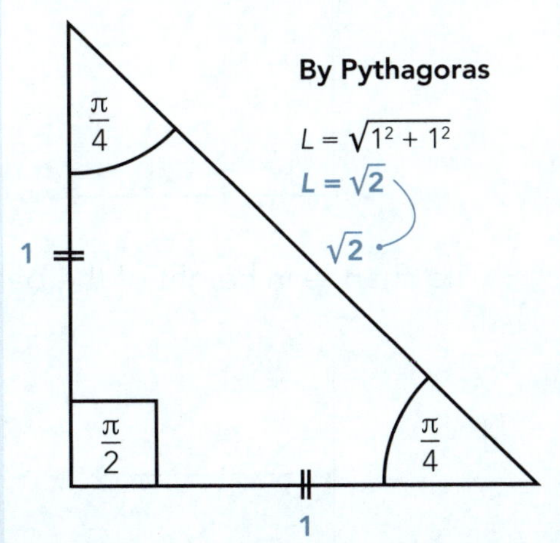

Using this triangle:

$$\sin\frac{\pi}{4} = \frac{1}{\sqrt{2}}$$

$$\cos\frac{\pi}{4} = \frac{1}{\sqrt{2}}$$

$$\tan\frac{\pi}{4} = 1$$

For $\frac{\pi}{6}$ and $\frac{\pi}{3}$ (or 30° and 60°), use half an equilateral triangle with sides 2 units long:

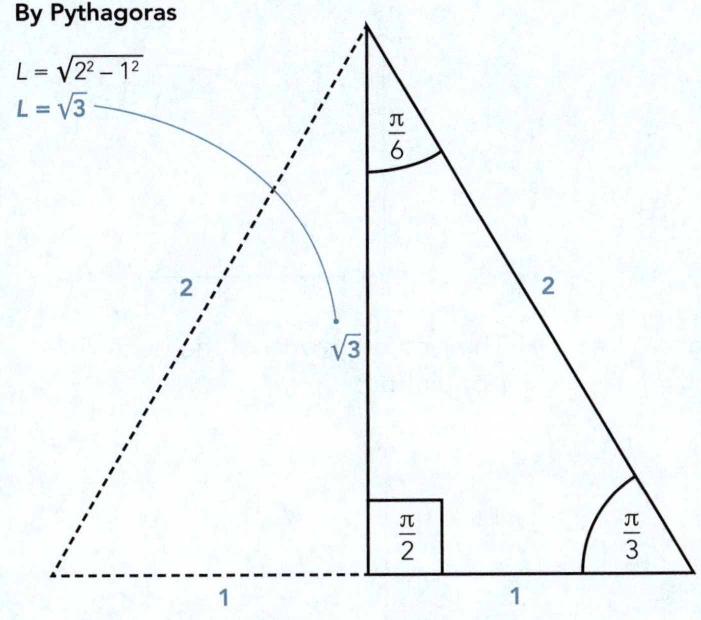

Using this triangle:

$$\sin\frac{\pi}{6} = \frac{1}{2} \qquad \sin\frac{\pi}{3} = \frac{\sqrt{3}}{2}$$

$$\cos\frac{\pi}{6} = \frac{\sqrt{3}}{2} \qquad \cos\frac{\pi}{3} = \frac{1}{2}$$

$$\tan\frac{\pi}{6} = \frac{1}{\sqrt{3}} \qquad \tan\frac{\pi}{3} = \sqrt{3}$$

Concepts

Calculus — the mathematics of change

- The gradient function shows how **one parameter is changing with respect to another**.

Examples:

1 The height of a plane during a steady climb as time changes:

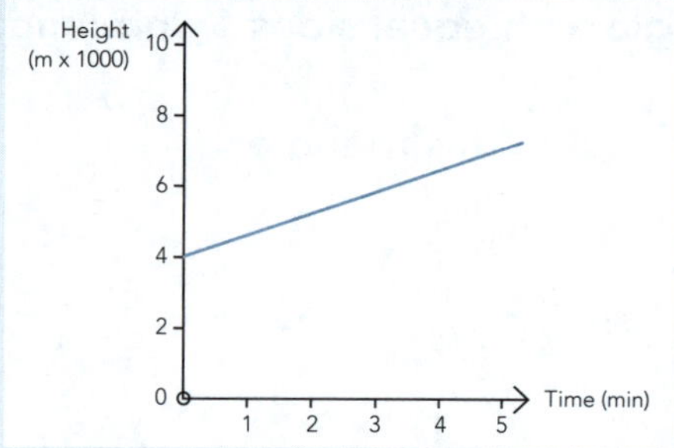

The change in height is 600 m/min:

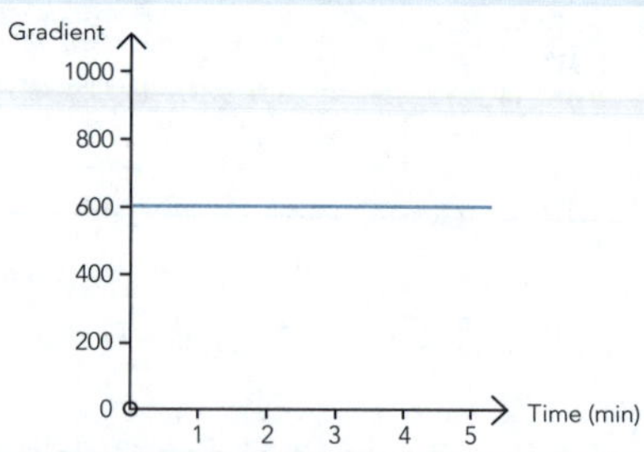

2 The trajectory of a ball thrown up from a balcony until it lands on the ground:

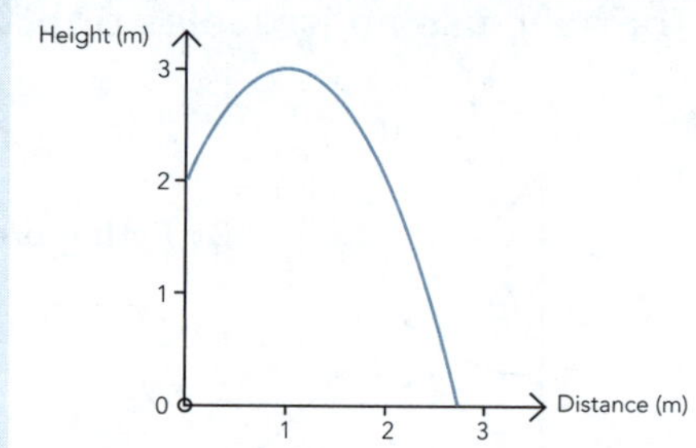

The change in height of the ball:

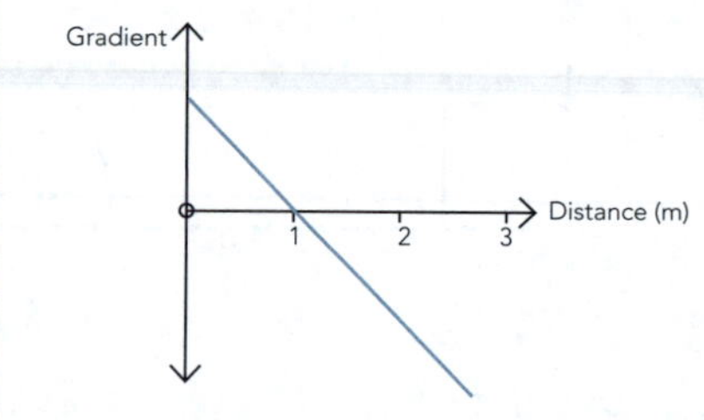

3 The height of a roller coaster above the ground:

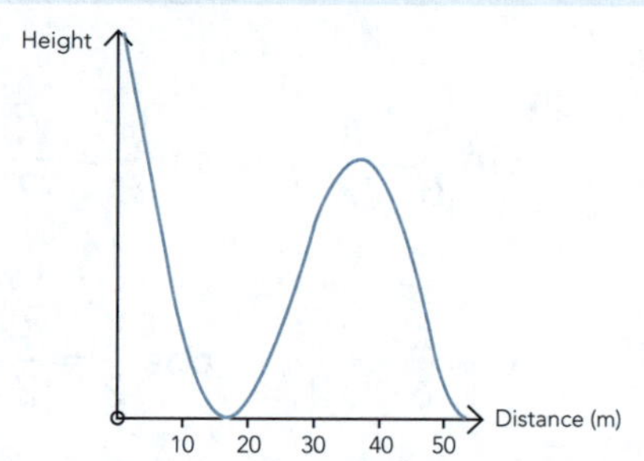

The change in height of the roller coaster:

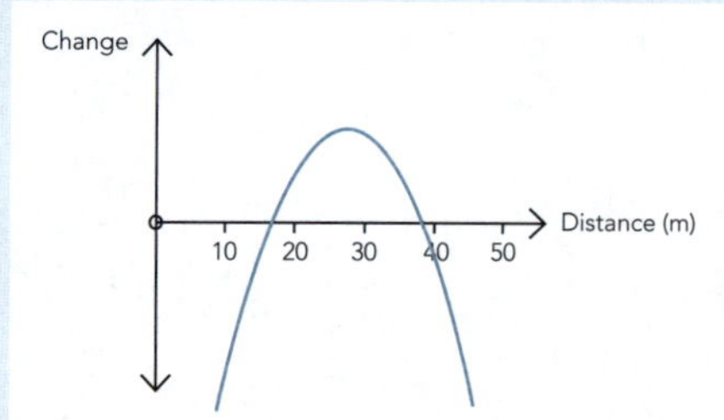

4 The number of bacteria in a test tube:

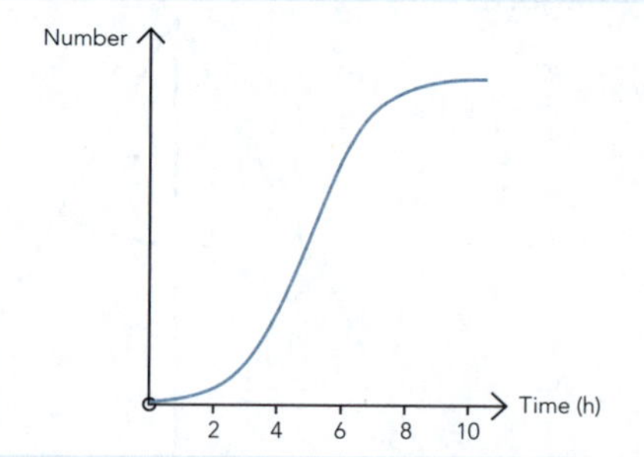

The rate of growth of the bacterial population:

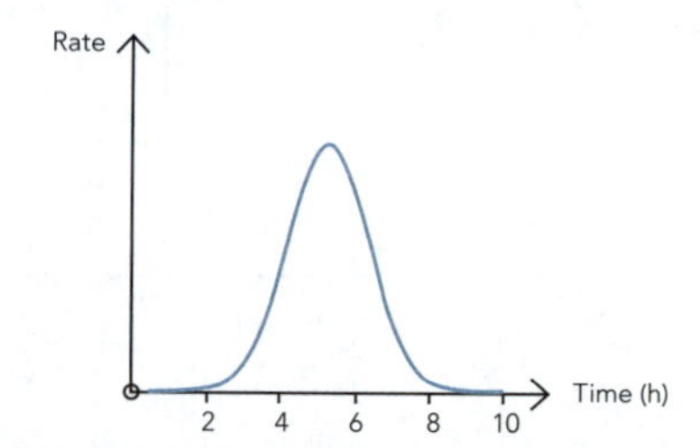

ISBN: 9780170446976

Sketching gradient functions

To discover the gradient function (derivative) and the second derivative of a function:

1. Sketch tangents at a range of points on the graph or gradient function, particularly where there is a turning point.
2. Write beside each point whether the gradient is positive, negative or zero.
3. Transfer these onto the graph of the gradient function ($f'(x)$ or $\frac{dy}{dx}$).
4. Repeat for the graph of the second derivative ($f''(x)$ or $\frac{d_2y}{dx^2}$).

Examples:

1 A quadratic: $y = -(x + 3)(x - 7)$

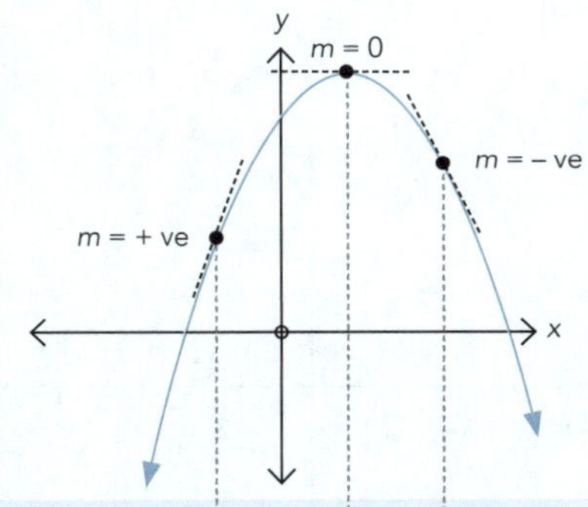

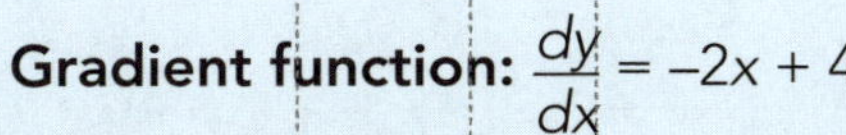

Gradient function: $\frac{dy}{dx} = -2x + 4$

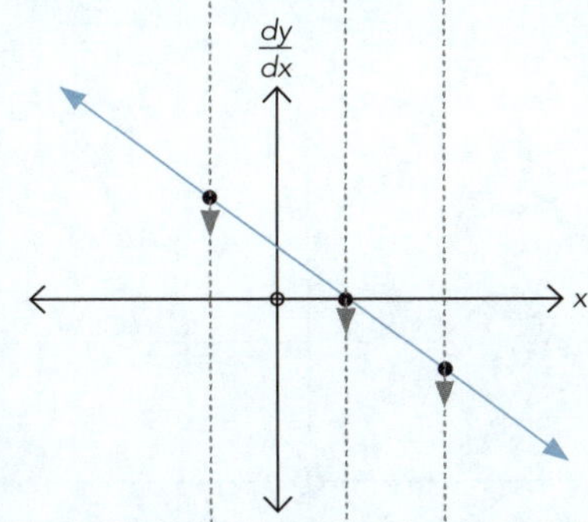

Second derivative: $\frac{d_2y}{dx^2} = -2$

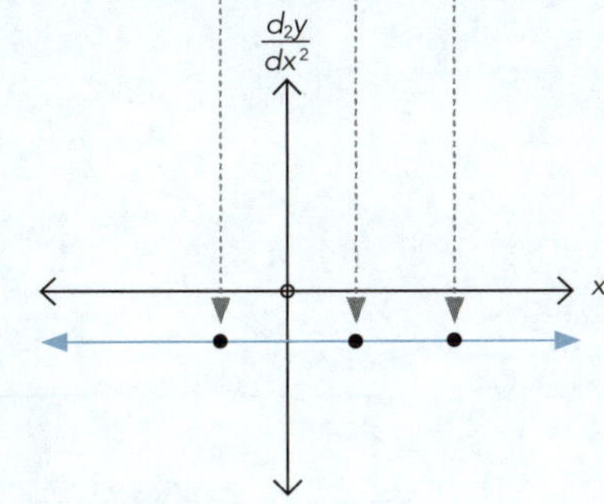

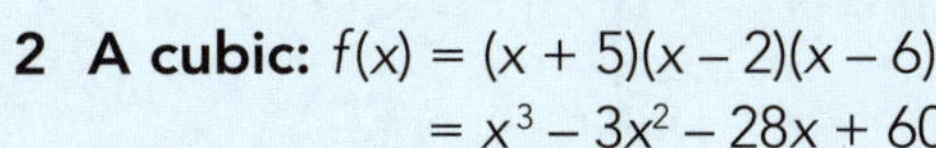

2 A cubic: $f(x) = (x + 5)(x - 2)(x - 6)$
$= x^3 - 3x^2 - 28x + 60$

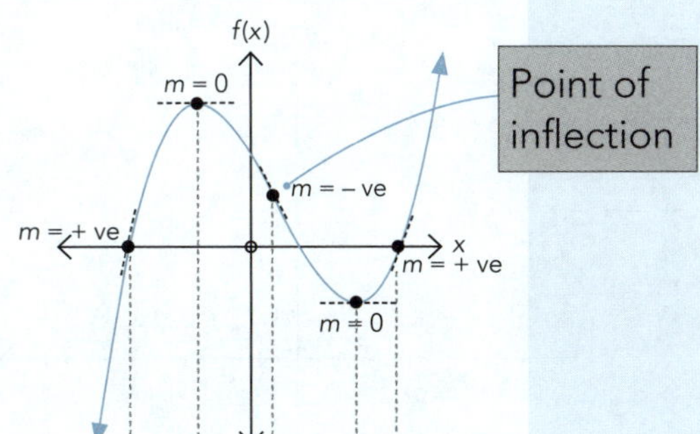

Gradient function: $f'(x) = 3x^2 - 6x - 28$

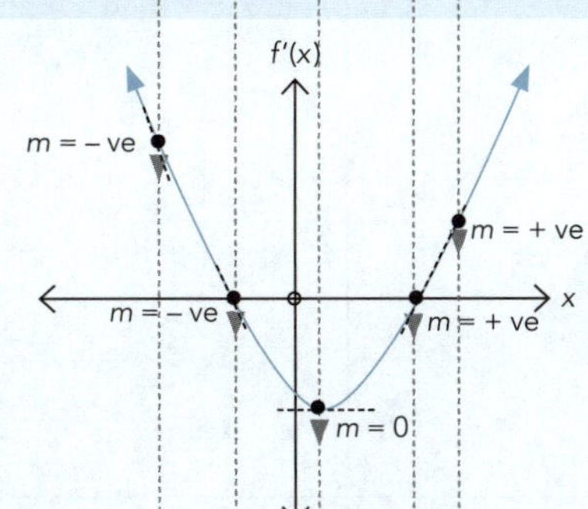

Second derivative: $f''(x) = 6x - 6$

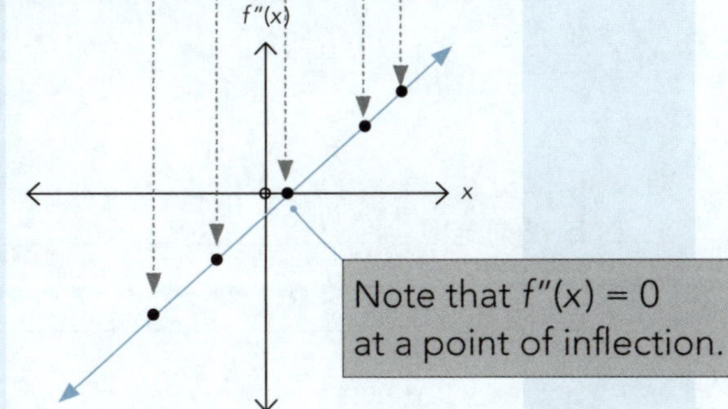

ISBN: 9780170446976

Sketch gradient and second derivative functions for the following. Then find the values of x (if any) that meet the conditions below.

1 a

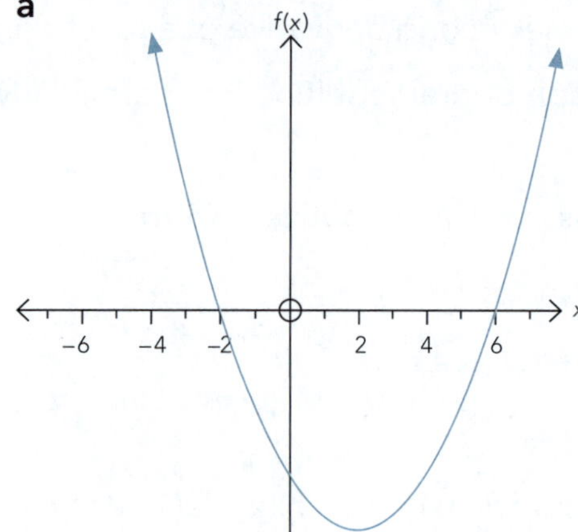

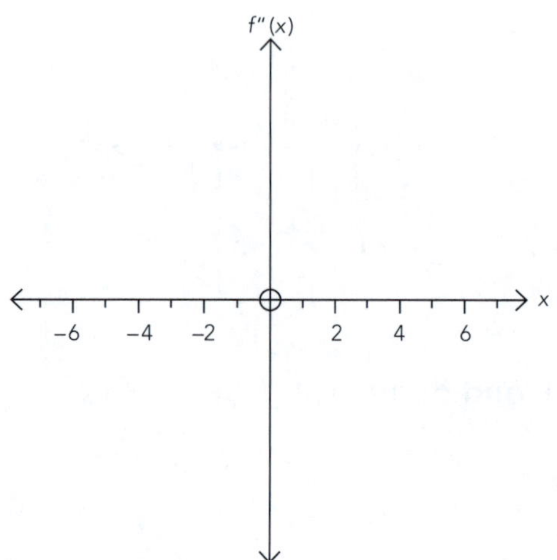

b $f(x) = 0$ ____________________

c $f'(x) = 0$ ____________________

d $f''(x) < 0$ ____________________

2 a

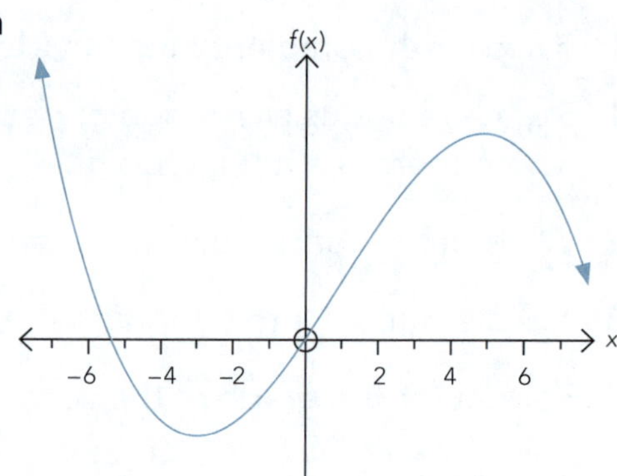

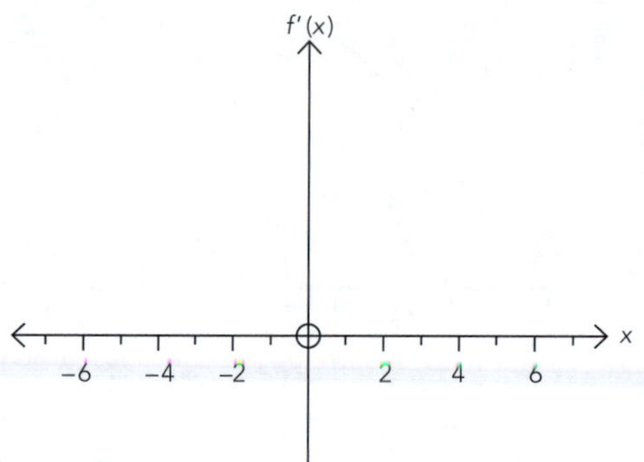

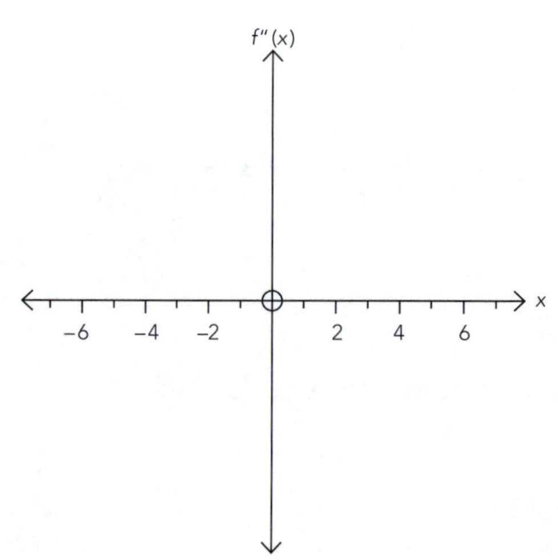

b $f'(x) = 0$ ____________________

c $f'(x) > 0$ ____________________

d $f''(x) < 0$ ____________________

ISBN: 9780170446976

3 **a**

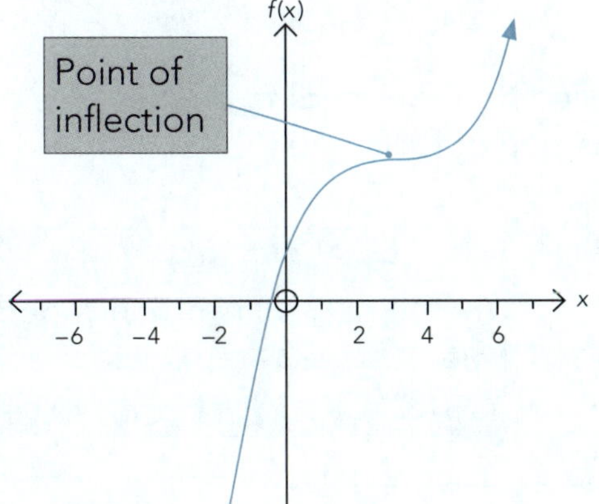

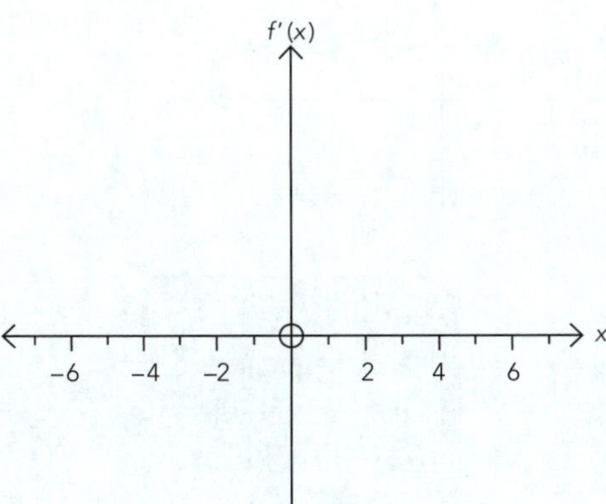

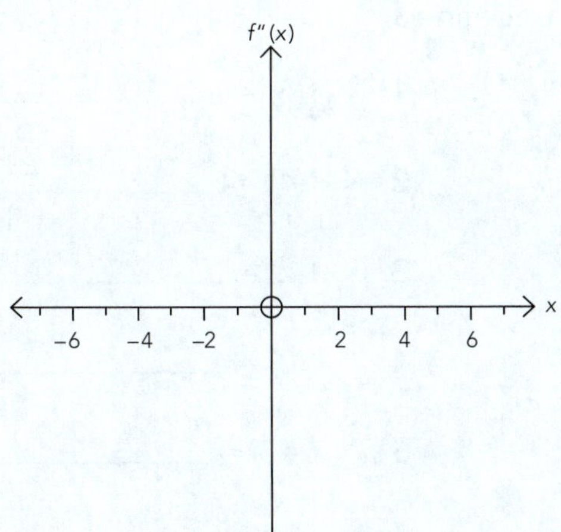

b $f'(x) = 0$ ________________

c $f'(x) < 0$ ________________

d $f''(x) \geq 0$ ________________

4 **a**

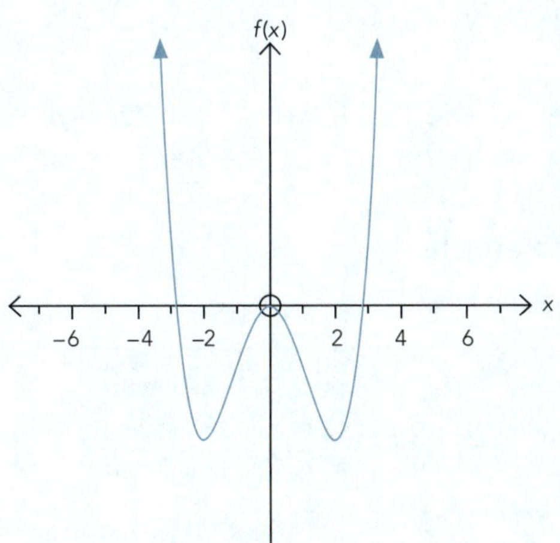

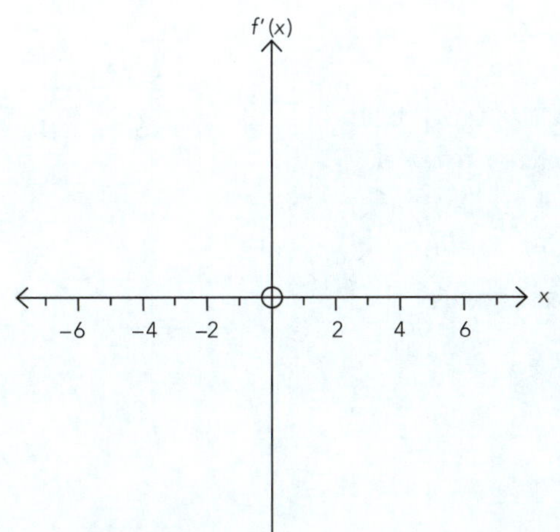

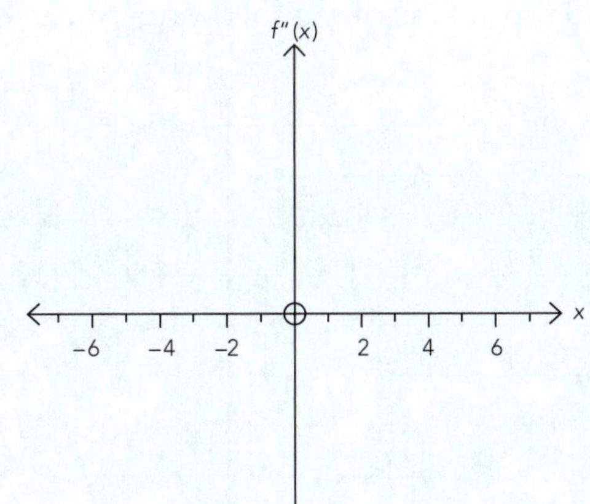

b $f'(x) = 0$ ________________

c $f'(x) > 0$ ________________

d $f''(x)$ has a negative gradient ________________

ISBN: 9780170446976

Piecewise functions

- These occur when there are different rules for finding y, depending on the value of x.
- A function is defined at a point if we know the value of $f(x)$.

Example:

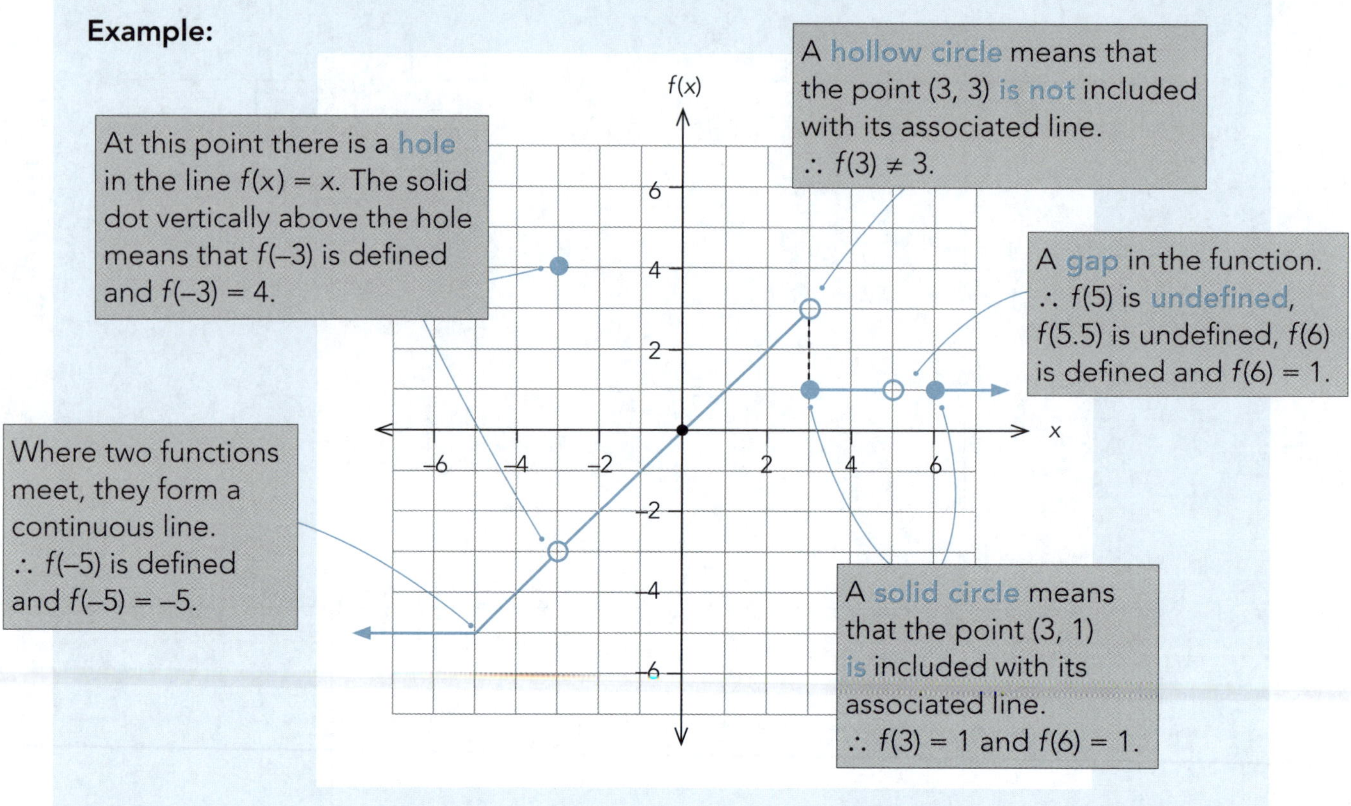

Find the values of the following. State if the value is undefined.

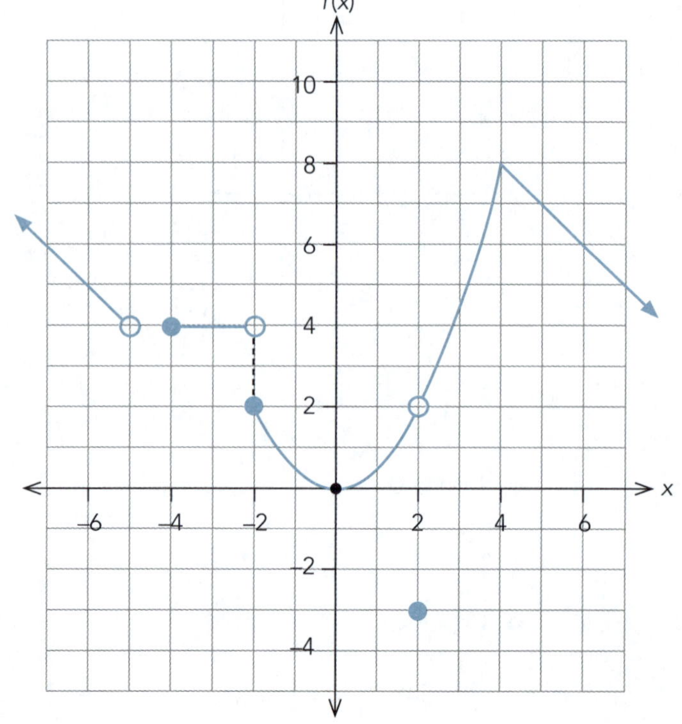

1 $f(6)$ = ______________

2 $f(4)$ = ______________

3 $f(2)$ = ______________

4 $f(0)$ = ______________

5 $f(-2)$ = ______________

6 $f(-3)$ = ______________

7 $f(-4)$ = ______________

8 $f(-4.5)$ = ______________

9 $f(-5)$ = ______________

10 $f(-6)$ = ______________

ISBN: 9780170446976

Limits

- A limit to a function at a point exists if both sides of the function approach the same value.
- The value of the limit of the function is the value of $f(x)$ that both parts approach.
- The value of the function at this point may **not** be the same as the value of the limit of the function. Consider a function with a hole:

$f(x) = \frac{1}{2}x$ with hole at $x = 4$, and point vertically above.

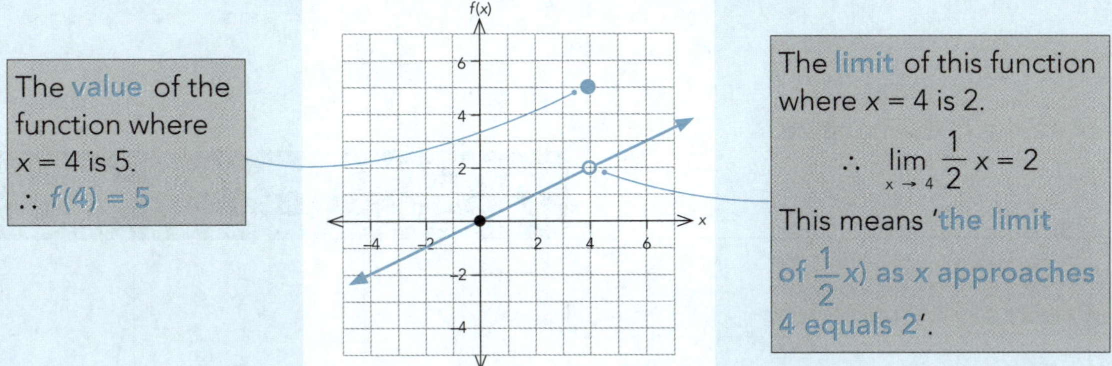

Examples where there is a limit:

1

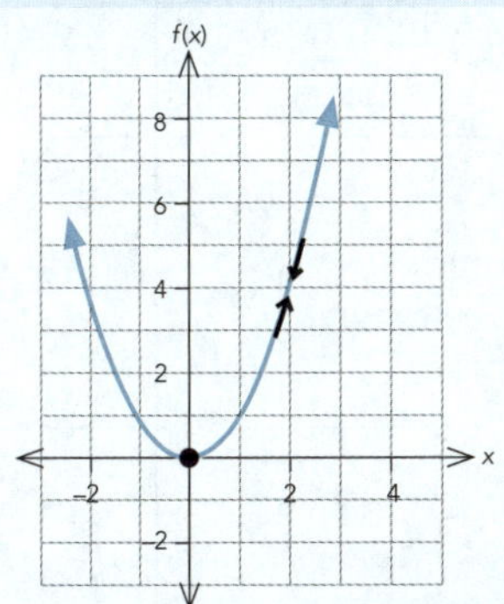

2

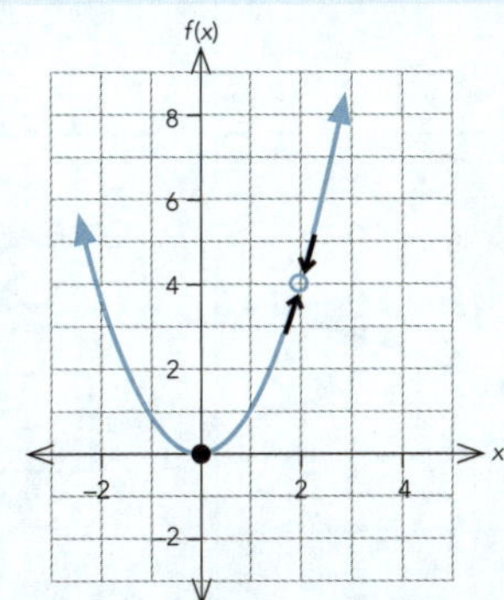

3

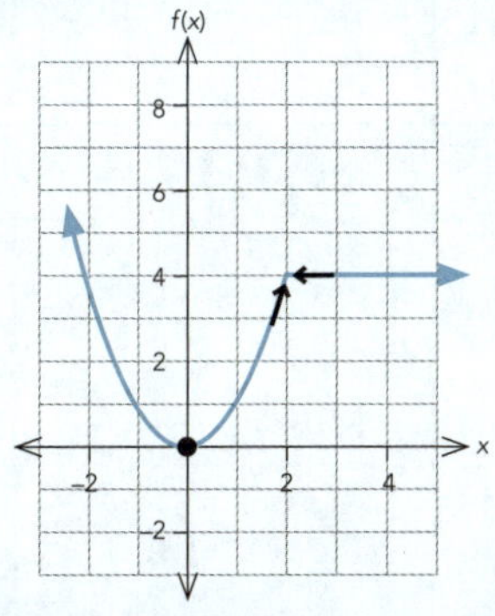

For all three of these, $\lim_{x \to 2} f(x) = 4$

Examples where a limit does not exist:

1

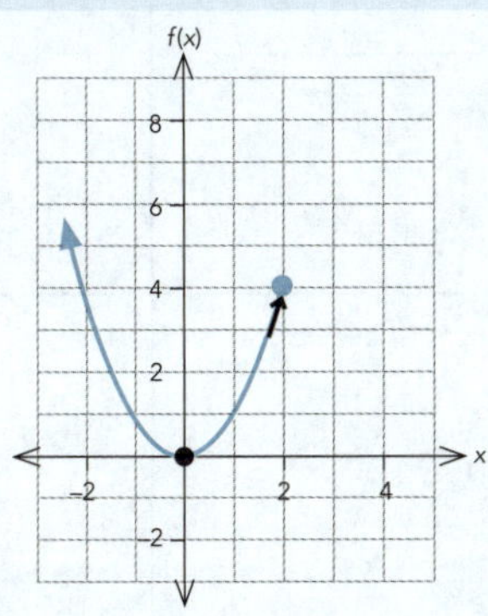

2

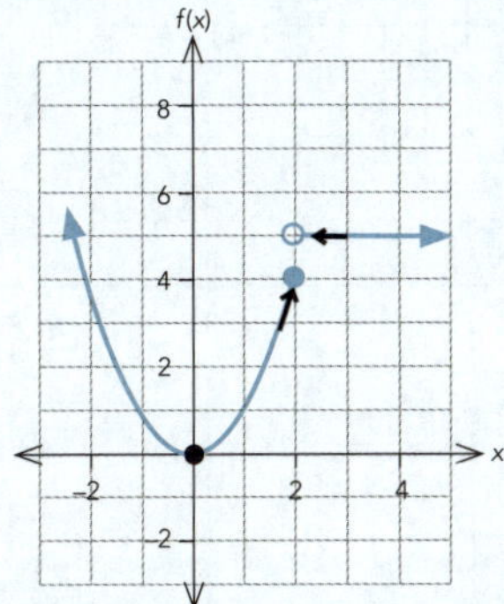

3

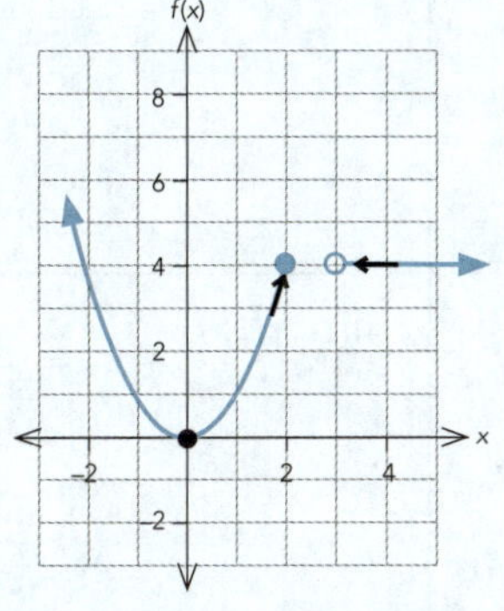

For all three of these, the limit where $x = 2$ **does not exist**.

ISBN: 9780170446976

4 Where there is a vertical asymptote:

Remember, an **asymptote** is a line that a curve approaches very closely, but never quite reaches. Because the asymptote is at $x = 2$, these functions are not defined where $x = 2$.

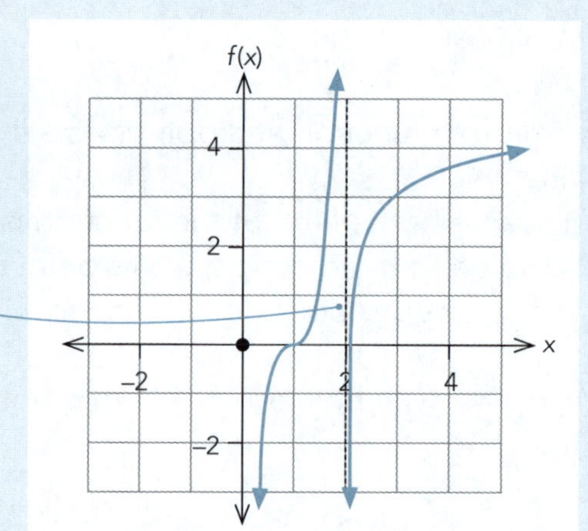

Complete the following table.

		Limit exists? (✓ or ×)	Value of limit if it exists	Value of $f(x)$ if it exists
1	$\lim_{x \to 2} f(x)$ 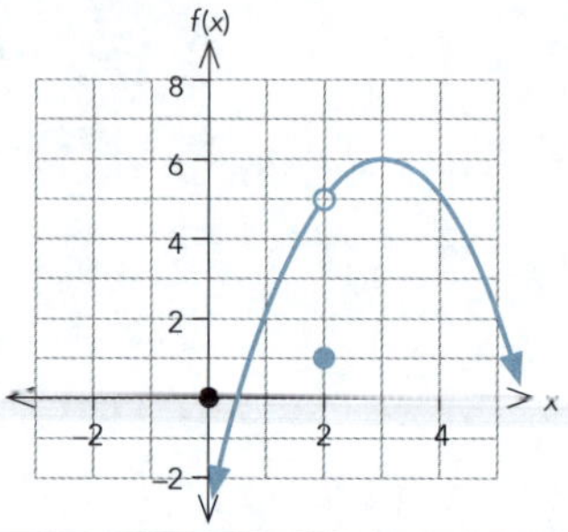	☐	______	______
2	$\lim_{x \to 0} f(x)$ 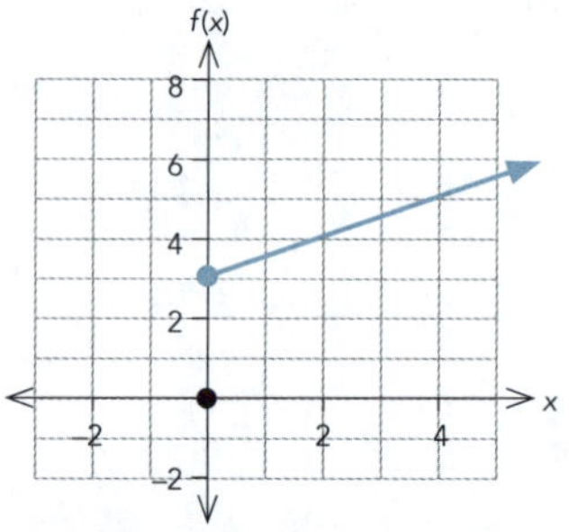	☐	______	______
3	$\lim_{x \to 3} f(x)$ 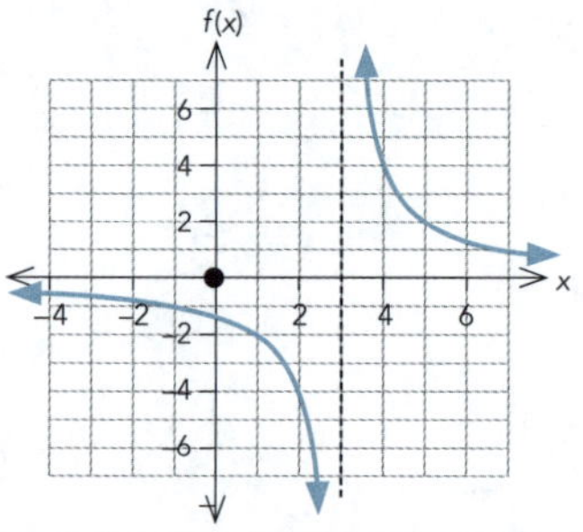	☐	______	______
4	$\lim_{x \to -3} f(x)$ 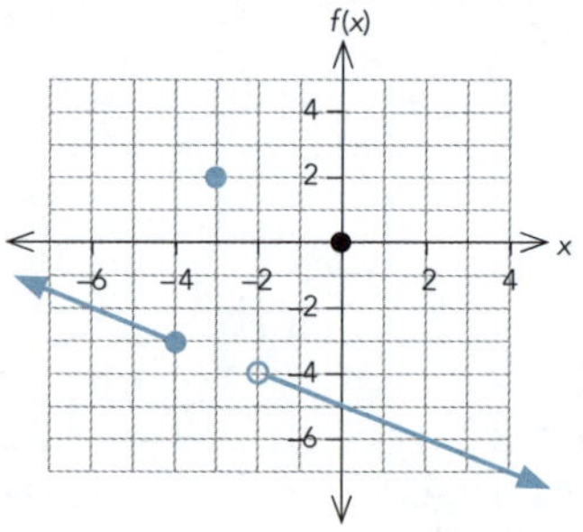	☐	______	______

ISBN: 9780170446976

Continuous functions

- A continuous function is one that could be drawn without lifting your pen from the paper.
- However, a function may be discontinuous at a point, but continuous everywhere else in its domain.

Examples of continuous functions:

1

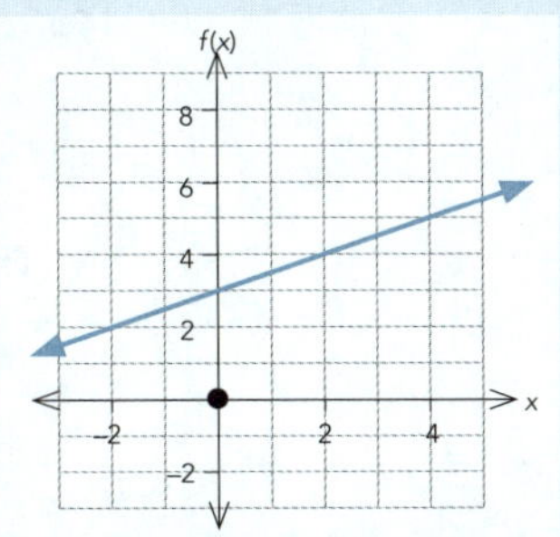

2

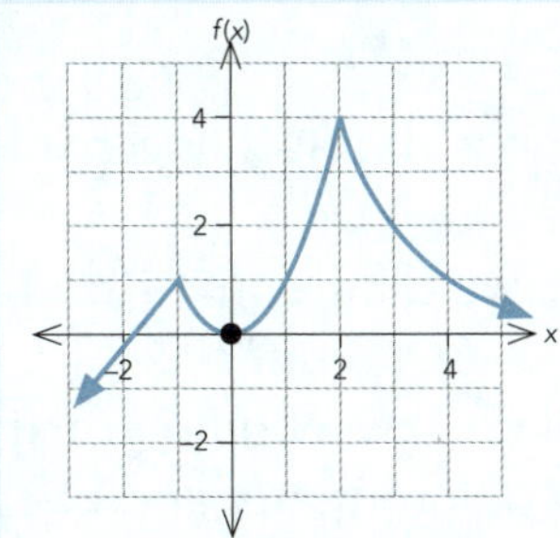

Examples of functions with discontinuities:

1

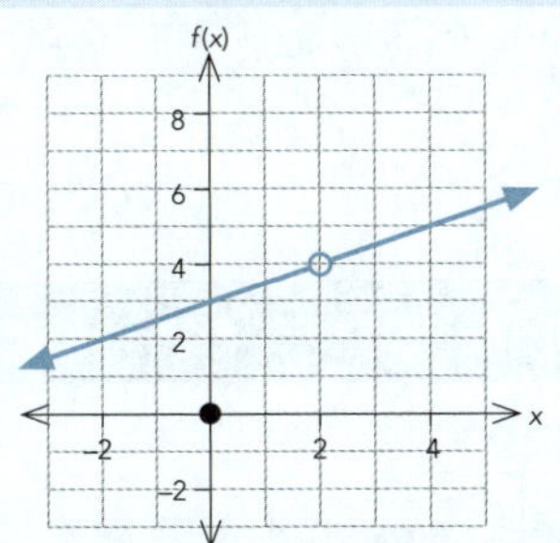

2

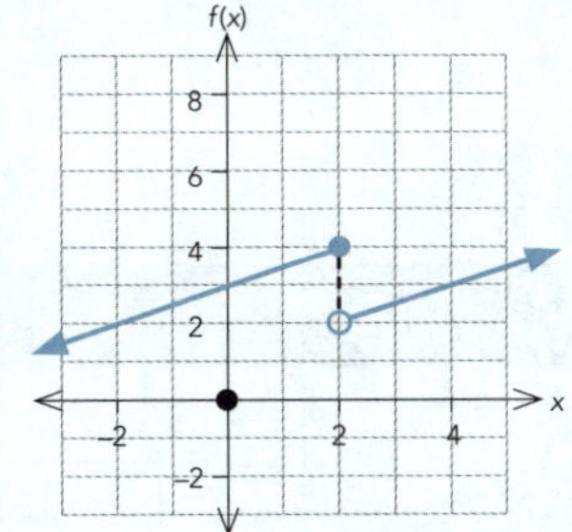

3

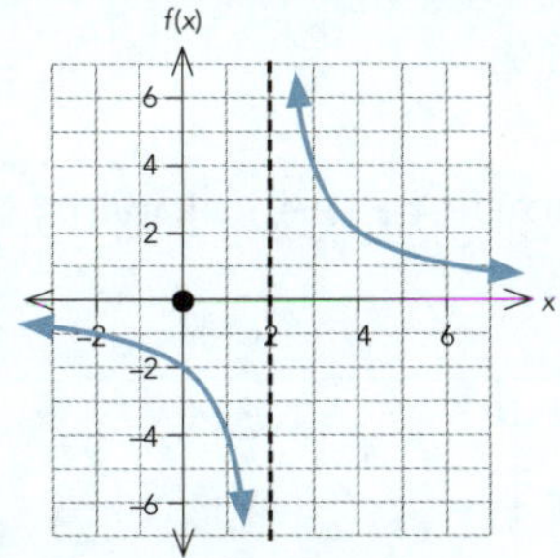

All three of these functions are discontinuous where $x = 2$, but continuous everywhere else.

State whether each of these functions is continuous or not (✓ or ✗) at these values for x.

1

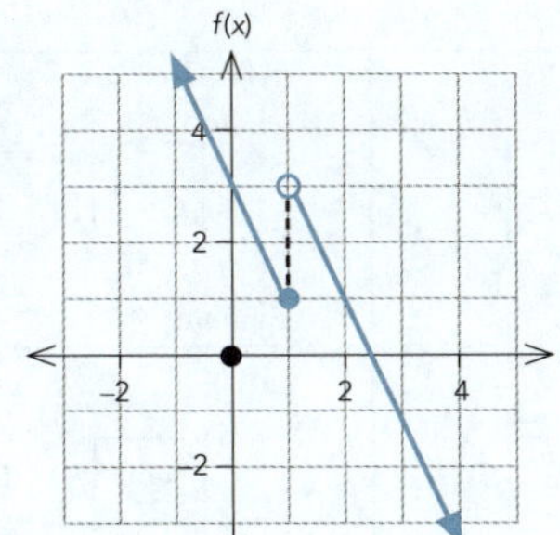

$x = 1$ 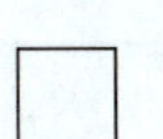

$x = 2$ ☐

2

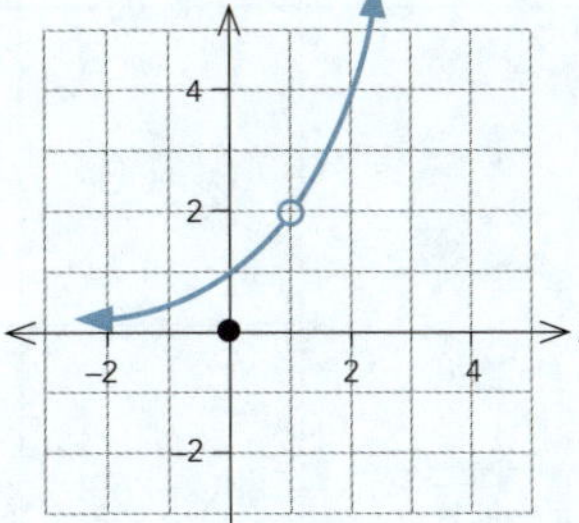

$x = 0$ ☐

$x = 1$ ☐

3

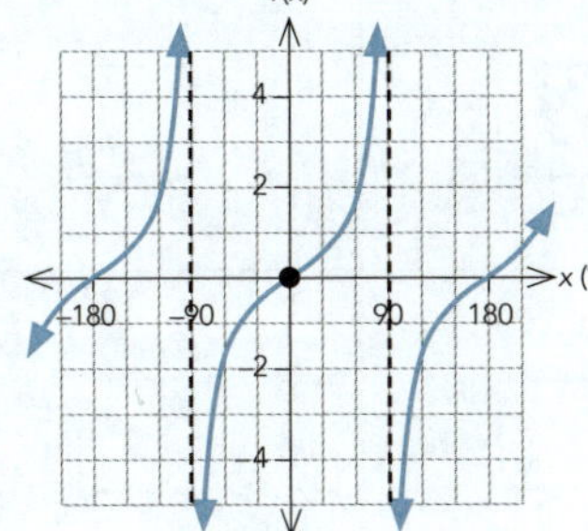

$x = 0°$ 

$x = 90°$ ☐

4

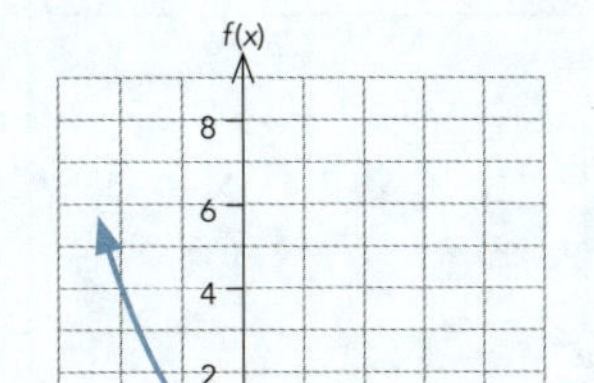

$x = -2$ ☐

$x = 1$ ☐

ISBN: 9780170446976

Differentiability

- In order to be differentiable at a point, a function must be **defined** and **continuous** at that point.
- Also, the slope of the tangent of points to the left must approach the same value as the slope of the tangent of points to the right.
- However, where two **different** functions meet, a function may be defined and continuous, but it will **not** be differentiable.

Example:

- This function is **defined** and **continuous** everywhere.
- Remember, we differentiate a function to find the **gradient function**.
- At (–1, 1) or (2, 4) we would not know which function to differentiate, and each would produce **different** gradient functions.
- So this function is differentiable everywhere **except** where $x = -1$ or $x = 2$.

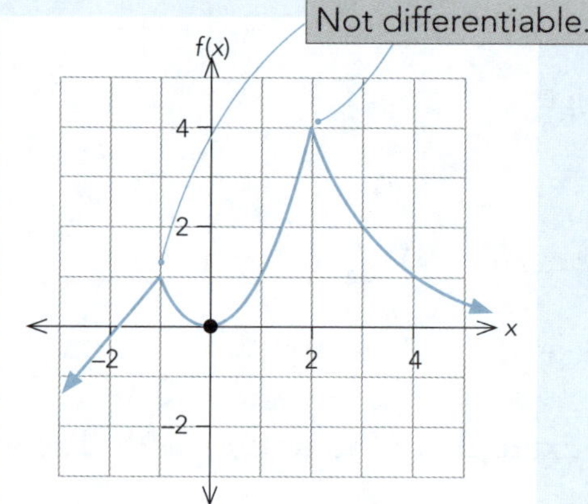

Complete the following table (✓ or ✗).

		x value	Defined?	Continuous?	Differentiable?
1	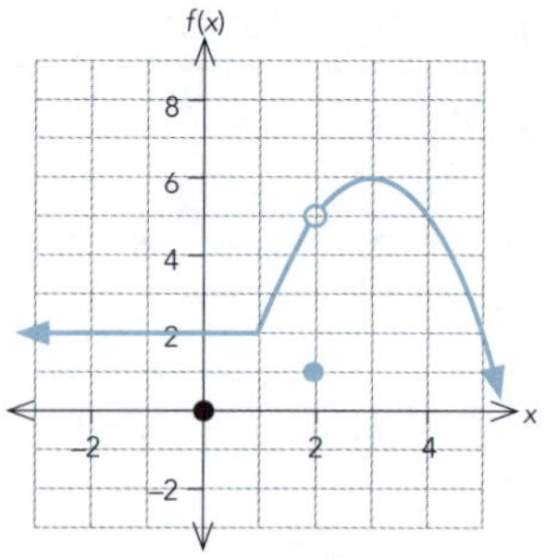	$x = 1$	☐	☐	☐
		$x = 2$	☐	☐	☐
2	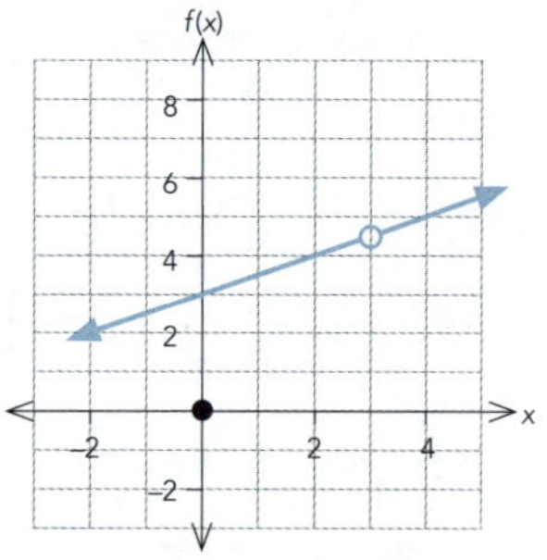	$x = 0$	☐	☐	☐
		$x = 3$	☐	☐	☐
3	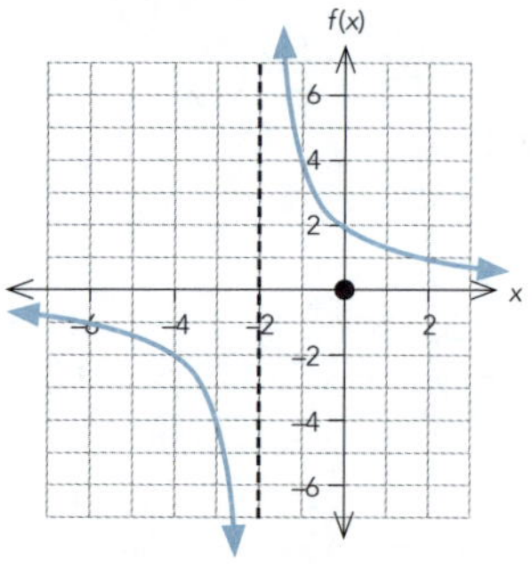	$x = 0$	☐	☐	☐
		$x = -2$	☐	☐	☐

  ISBN: 9780170446976

Putting it all together

1 Complete the following table.

	Defined at x = 2?	f(2)	Limit exists?	$\lim_{x \to 2}$	Continuous at x = 2?	Differentiable at x = 2?
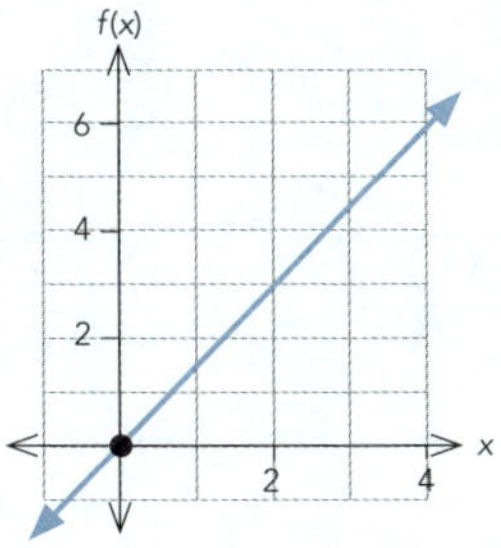			Exists			
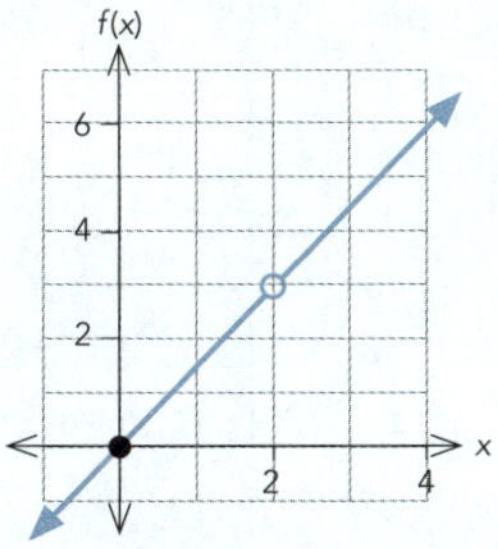						✗
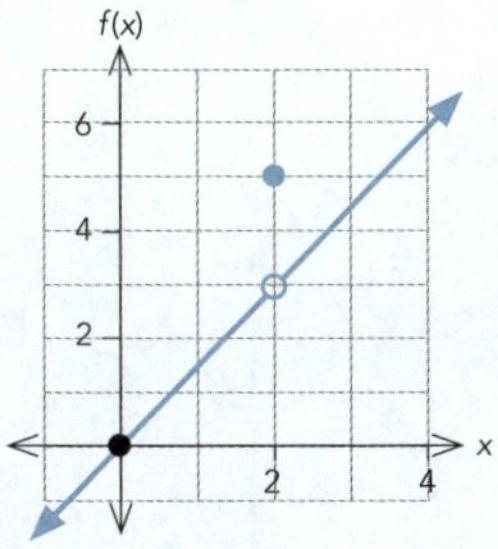	✓					
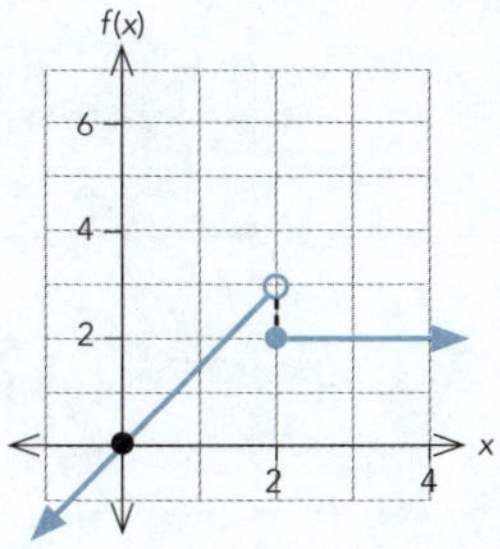					✗	

ISBN: 9780170446976

	Defined at x = 2?	$f(2)$	Limit exists?	$\lim\limits_{x \to 2}$	Continuous at $x = 2$?	Differentiable at $x = 2$?
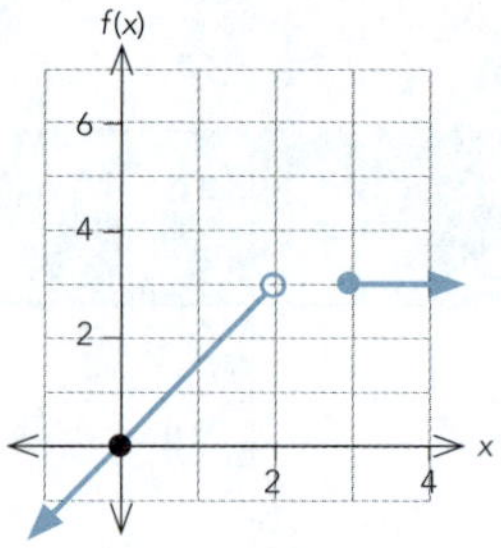		Undefined				
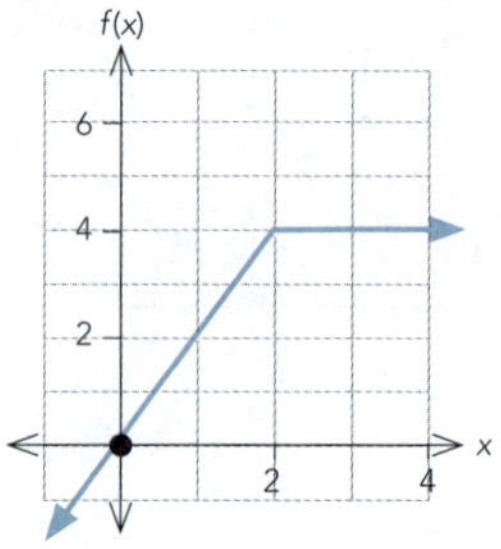				4		
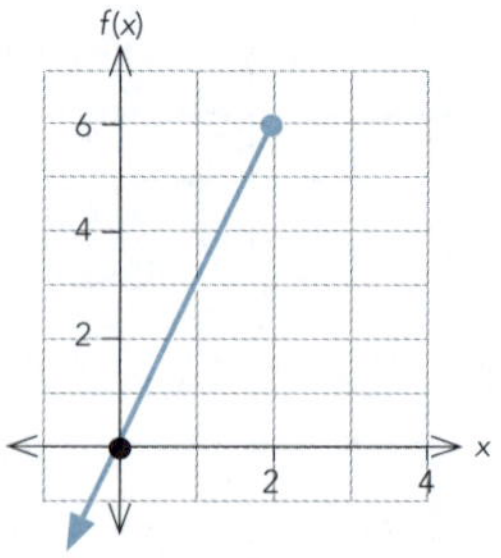						
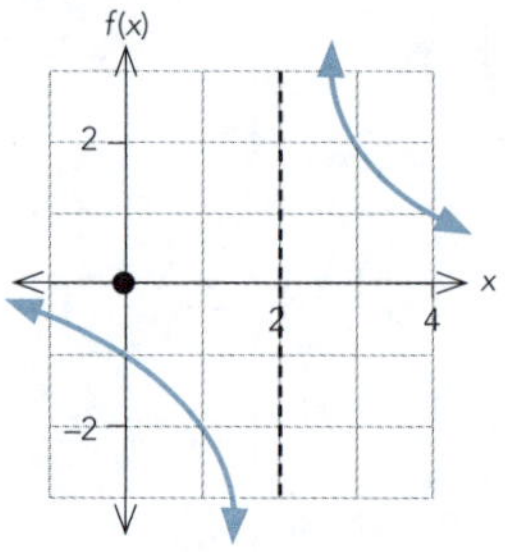						

ISBN: 9780170446976

2

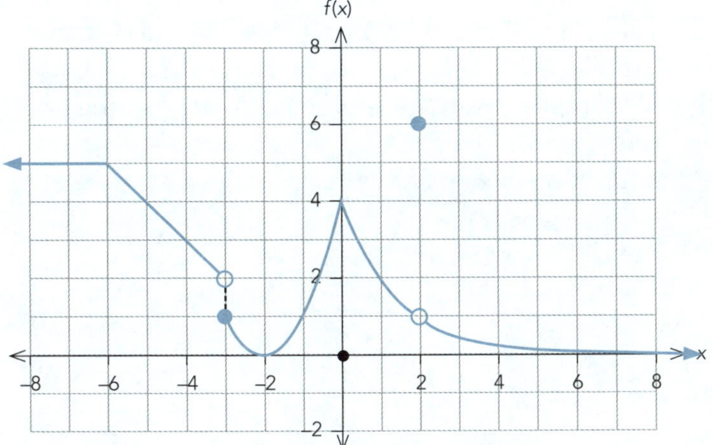

a For this function, find the value(s) of x that meet the following conditions.

i $f'(x) = 0$ ________________

ii $f(x)$ is not continuous

____

iii $f'(x) > 0$ ________________

iv $f(x)$ is continuous but not differentiable

____

b $f(-3) =$ ________________

c What is the value of $\lim_{x \to 2} f(x)$?

State clearly if this does not exist.

____

3

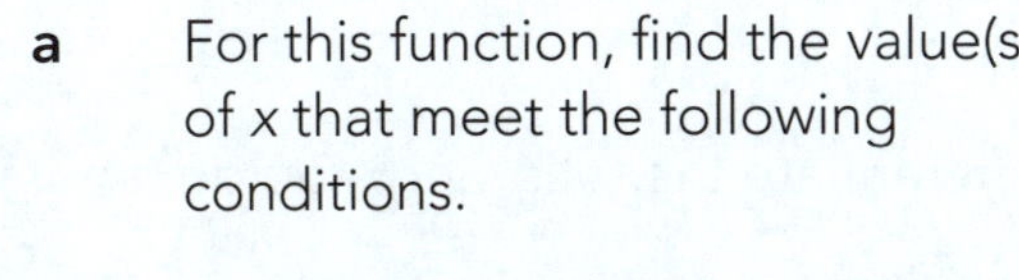

a For this function, find the value(s) of x that meet the following conditions.

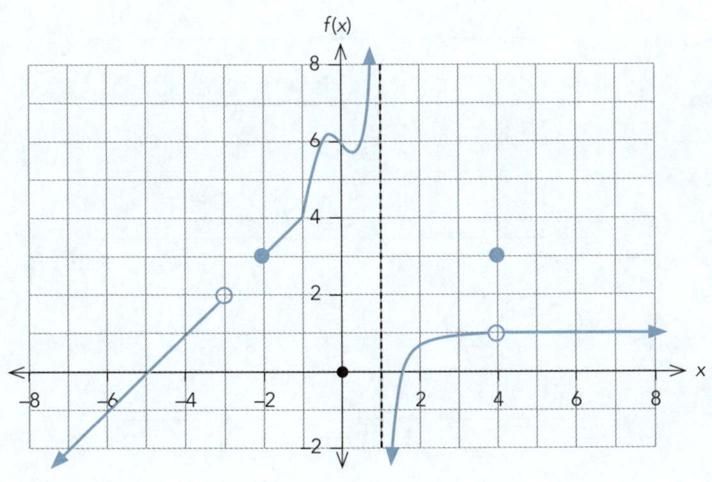

i $f(x)$ is not defined

____

ii $f'(x)$ is a constant

____

iii $f(x)$ is continuous but not differentiable

____

b $f(4) =$ ________________

c What is the value of $\lim_{x \to 4} f(x)$?

State clearly if this does not exist.

____

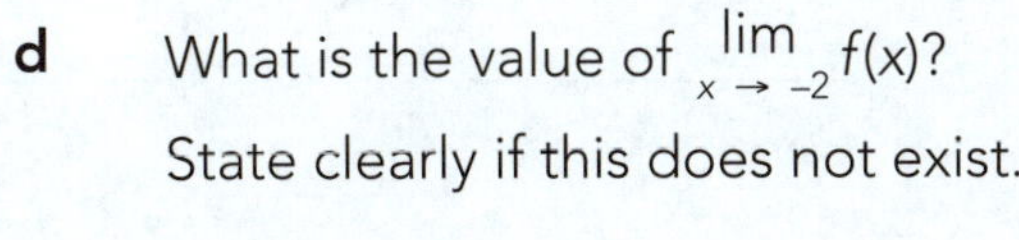

d What is the value of $\lim_{x \to -2} f(x)$?

State clearly if this does not exist.

____

ISBN: 9780170446976

Differentiation techniques

1 Expressions with fractional and negative indices

- Apply the **same rules** as you would to polynomials, being careful to correctly use the exponent laws.
- Before you start to differentiate, write the function in **exponent form**.

Remember: $\frac{a}{x^n} = ax^{-n}$ and $\sqrt[n]{x^m} = x^{\frac{m}{n}}$

Examples:

1 $y = \frac{5}{x^3} = 5x^{-3}$

$\frac{dy}{dx} = -15x^{-4}$

$= \frac{-15}{x^4}$

2 $f(x) = 12\sqrt[3]{x} = 12\, x^{\frac{1}{3}}$

$f'(x) = \frac{1}{3} \cdot 12x^{-\frac{2}{3}}$

$= 4 \cdot \frac{1}{\sqrt[3]{x^2}}$

$= \frac{4}{\sqrt[3]{x^2}}$

This is all that will be required in your examination, but it is useful to be able to simplify it.

Differentiate the following. You do not have to simplify your answers.

1 $f(x) = \frac{1}{x^4}$

2 $y = 6\sqrt{x}$

3 $y = 5\sqrt[3]{x}$

4 $f(x) = \frac{1}{2x}$

 ISBN: 9780170446976

5 $f(x) = \dfrac{2}{x^5}$

6 $y = \dfrac{1}{\sqrt{x}}$

7 $y = \dfrac{3}{4x^2}$

8 $f(x) = 6\sqrt[3]{x^2}$

9 $f(x) = x\sqrt{x}$

10 $y = \dfrac{2x^3}{5\sqrt{x}}$

11 $y = \dfrac{8}{3\sqrt[4]{x}}$

12 $f(x) = \dfrac{4}{3x^2\sqrt{x}}$

ISBN: 9780170446976

2 The chain rule

- This is used where there is a function of a function: $f(g(x))$
- Both of these forms of reasoning are on your formula sheet:

$$(f(g))' = f'(g).g' \quad \text{or} \quad \text{if } y = f(u) \text{ and } u = g(x), \text{ then } \frac{dy}{dx} = \frac{dy}{du} \times \frac{du}{dx}$$

Notice: To make expressions less cumbersome, we drop the '(x)' from functions.

Notice: A dot (.) is commonly used instead of a multiplication sign.

These are on your formula sheet.

Examples:

1 Function notation

$f(x) = \sqrt{3x^2 - 2}$

Let $f = (g)^{\frac{1}{2}}$ $\quad g = 3x^2 - 2$

Then $f' = \frac{1}{2}g^{-\frac{1}{2}}$ $\quad g' = 6x$

But $f'(x) = f'(g) . g'$

so $f'(x) = \frac{1}{2}g^{\frac{1}{2}} . 6x$

$= \frac{1}{2}(3x^2 - 2)^{-\frac{1}{2}} . 6x$

$\therefore f'(x) = \frac{3x}{\sqrt{3x^2 - 2}}$

Leibniz notation

$y = \sqrt{3x^2 - 2} = (3x^2 - 2)^{\frac{1}{2}}$

Let $y = u^{\frac{1}{2}}$ $\quad u = (3x^2 - 2)$

Then $\frac{dy}{du} = \frac{1}{2}u^{-\frac{1}{2}}$ $\quad \frac{du}{dx} = 6x$

$= \frac{1}{2\sqrt{u}}$

But $\frac{dy}{dx} = \frac{dy}{du} \times \frac{du}{dx}$

so $= \frac{1}{2\sqrt{3x^2 - 2}} . 6x$

$\frac{dy}{dx} = \frac{3}{\sqrt{3x^2 - 2}}$

2 Function notation

$f(x) = 2(4x^2 + 5)^3 = f(g)$

Let $f = 2g^3$ $\quad g = 4x^2 + 5$

Then $f' = 6g^2$ $\quad g' = 8x$

But $f'(x) = f'(g) . g'$

so $= 6g^2 . 8x$

$= 6(4x^2 + 5)^2 . 8x$

$\therefore f'(x) = 768x^5 + 1920x^3 + 1200x$

Leibniz notation

$y = 2(4x^2 + 5)^3$

Let $y = 2u^3$ $\quad u = 4x^2 + 5$

Then $\frac{dy}{du} = 6u^2$ $\quad \frac{du}{dx} = 8x$

But $\frac{dy}{dx} = \frac{dy}{du} \times \frac{du}{dx}$

so $\frac{dy}{dx} = 6u^2 . 8x$

$= 6(4x^2 + 5)^2 . 8x$

$\therefore \frac{dy}{dx} = 768x^5 + 1920x^3 + 1200x$

ISBN: 9780170446976

Use the chain rule to differentiate the following.

1 $f(x) = (x + 4)^2$

2 $y = 4(x^2 - 3)^3$

3 $y = 5(8 - 3x^2)^2$

4 $f(x) = 6(x^2 - 2x)^4$

5 $f(x) = (2x^3 - x + 10)^4$

6 $y = (x^2 + 3x + 4)^7$

ISBN: 9780170446976

7 $y = \sqrt{5x + 1}$

8 $f(x) = \frac{1}{2x - 1}$

9 $f(x) = \frac{1}{(5x + 1)^2}$

10 $y = \frac{4}{(x - 3)^2}$

11 $y = \frac{6}{\sqrt{3 - x}}$

12 $f(x) = \sqrt{2x^4 - 5}$

ISBN: 9780170446976

3 Exponential functions

- The exponential function is written **exp(x)** or e^x for short.
- *e* stands for Euler's (pronounced 'oilers') number, and it equals **2.718 281 828 …**
- Its unique property is that the **gradient at any point is the same as its *y* value**.
- This means that **e^x has itself as its derived function.**

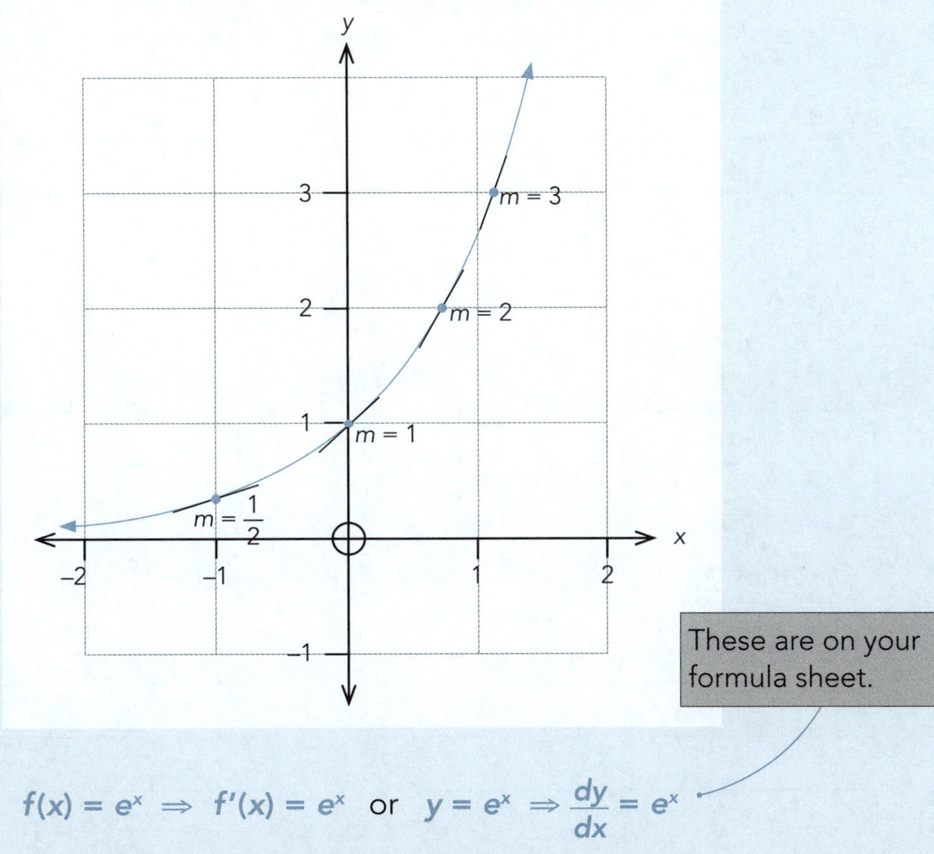

These are on your formula sheet.

$$f(x) = e^x \Rightarrow f'(x) = e^x \quad \text{or} \quad y = e^x \Rightarrow \frac{dy}{dx} = e^x$$

- Almost all examples needing differentiation of exponential functions involve use of the **chain rule**, and often the **product or quotient rule**.

Example:

Function notation

$f(x) = 7e^{3x^2 + 5x}$

$f = 7e^g$

$g = 3x^2 + 5x$

$f'(g) = 7e^g$

$g' = 6x + 5$

$\therefore f'(g).g' = 7e^g(6x + 5)$

$f'(x) = 7e^{3x^2 + 5x}.(6x + 5)$

$= 7(6x + 5)e^{3x^2 + 5x}$

Leibniz notation

$y = 7e^{3x^2 + 5x}$

Use the chain rule.

Let $y = 7e^u$ $\quad \therefore u = 3x^2 + 5x$

Then $\frac{dy}{du} = 7e^u$ $\quad \frac{du}{dx} = 6x + 5$

$\therefore \frac{dy}{dx} = 7e^u.(6x + 5)$

$\frac{dy}{dx} = 7(6x + 5)e^{3x^2 + 5x}$

ISBN: 9780170446976

Differentiate the following using the chain rule.

1 $y = e^{5x}$

2 $f(x) = e^{-x}$

3 $f(x) = e^{x-3}$

4 $y = e^{x^2}$

5 $y = e^{2-7x}$

6 $f(x) = 4e^{3x}$

7 $f(x) = 10e^{x^2-3}$

8 $y = 2e^{3x^2+x}$

9 $f(x) = e^{\frac{1}{x^2}}$

10 $y = \frac{1}{e^{2x}}$

11 $y = 3e^{5x} + x - 4$

12 $f(x) = 2e^{(x-4)^2}$

ISBN: 9780170446976

4 Logarithmic functions

- The logarithmic function is written **ln(x)**, which means **$\log_e(x)$**.
- Remember, **e** stands for Euler's number, and it equals **2.718 281 828 …**
- The graph of $y = \ln(x)$ is the **inverse** of the graph of $y = e^x$, which means it is a reflection in the line $y = x$.
- The gradient at any point on the graph of **ln(x)** is $\dfrac{1}{\text{the x value at that point}}$.

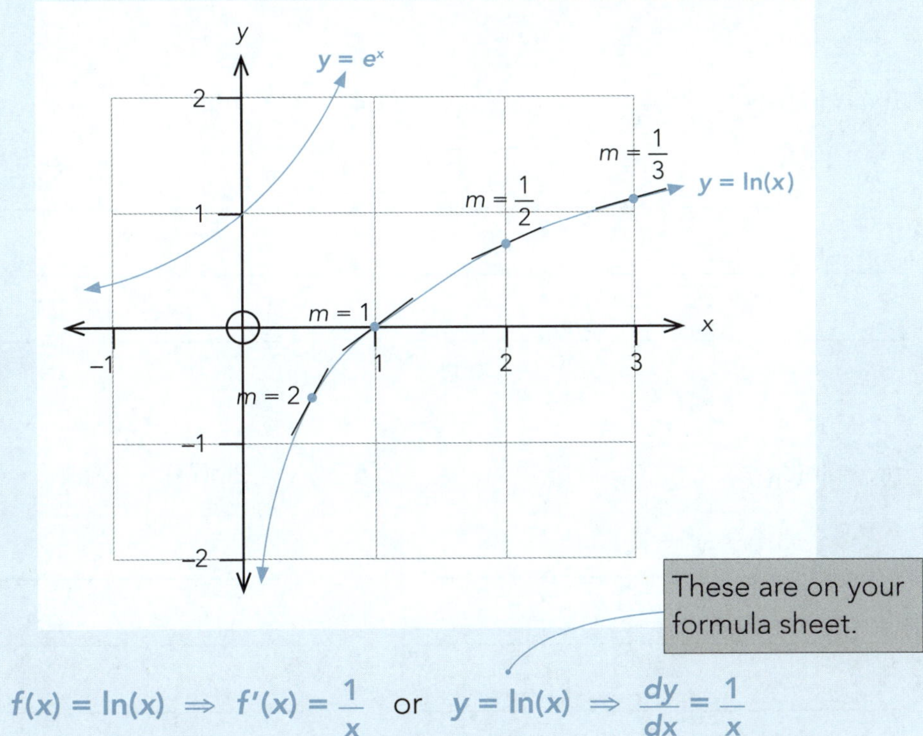

These are on your formula sheet.

$$f(x) = \ln(x) \Rightarrow f'(x) = \frac{1}{x} \quad \text{or} \quad y = \ln(x) \Rightarrow \frac{dy}{dx} = \frac{1}{x}$$

- As with exponential functions, almost all examples needing differentiation of logarithmic functions involve use of the **chain rule**, and often the **product rule**.

Examples:

1 Function notation

$$f(x) = \ln(2x^3)$$
$$f = \ln(g)$$
$$g = 2x^3$$
$$f'(g).g' = \frac{1}{g}.6x^2$$
$$\therefore f'(x) = \frac{6x^2}{2x^3}$$
$$= \frac{3}{x}$$

2 Leibniz notation

$$y = \ln(5x - e^x)$$

Let $y = \ln u$ $\quad \therefore u = 5x - e^x$

Then $\dfrac{dy}{du} = \dfrac{1}{u}$ $\quad \dfrac{du}{dx} = 5 - e^x$

Use the chain rule.

$$\therefore \frac{dy}{dx} = \frac{dy}{du}.\frac{du}{dx}$$
$$= \frac{1}{u}.(5 - e^x)$$
$$\frac{dy}{dx} = \frac{5 - e^x}{5x - e^x}$$

Differentiate the following.

1 $y = \ln(6x)$

2 $f(x) = \ln(7x + 2)$

3 $f(x) = \ln(5x) - 3$

4 $y = \ln(x^4)$

5 $f(x) = \ln(3x^5)$

6 $y = 2\ln(7x)$

7 $y = 4\ln(x^2 - 1)$

8 $f(x) = \ln(2x^3 + x)$

9 $f(x) = 7x^2 + 2\ln(x)$

10 $y = 5\ln(9x - 2)$

ISBN: 9780170446976

5 Trigonometric functions

Trigonometric functions other than sine, cosine and tangent

- You are familiar with sine, cosine and tangent functions.
- There are three new functions that you must become familiar with:

    cosec θ (cosecant)
    sec θ (secant)
    and cot θ (cotangent).
- **In order to differentiate trigonometric functions, angles must be in radians.**

cosecant: $y = \text{cosec } \theta$

In a right-angled triangle, **cosec** $\theta = \dfrac{\text{hypotenuse}}{\text{opposite}} = \dfrac{1}{\sin\theta}$

secant: $y = \sec\theta$

In a right-angled triangle, **sec** $\theta = \dfrac{\text{hypotenuse}}{\text{adjacent}} = \dfrac{1}{\cos\theta}$

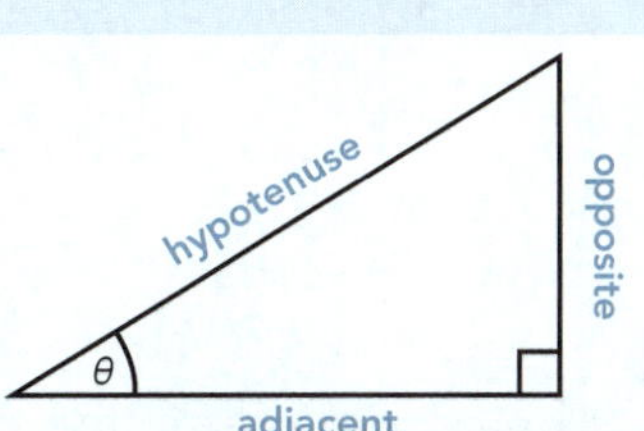

cotangent: $y = \cot\theta$

In a right-angled triangle, **cot** $\theta = \dfrac{\text{adjacent}}{\text{opposite}} = \dfrac{1}{\tan\theta}$

New trigonometric identities:

These are on your formula sheet.

$\text{cosec } \theta = \dfrac{1}{\sin\theta}$	$\sec\theta = \dfrac{1}{\cos\theta}$	$\cot\theta = \dfrac{1}{\tan\theta}$

- These are useful when differentiating functions.

Other useful relationships: $\dfrac{\sin(x)}{\cos(x)} = \tan(x)$ and $\dfrac{\cos(x)}{\sin(x)} = \cot(x)$

Examples: Write the following without using fractions.

1 $y = \dfrac{5}{\sin(x)} \Rightarrow y = 5\text{cosec}(x)$

2 $f(x) = \dfrac{\cos(2x)}{\sin(2x)} \Rightarrow f(x) = \cot(2x)$

Write the following without using fractions.

1 $y = \dfrac{2}{\cos(x)}$ ________________

2 $f(x) = \dfrac{x}{\tan(3x)}$ ________________

3 $f(x) = \dfrac{3\sin(6x)}{3\cos(6x)}$ ________________

4 $y = \dfrac{9}{\sec(x)}$ ________________

ISBN: 9780170446976

Differentiation of trigonometric functions

- You can tell from their graphs that:

1 The gradient function of $y = \sin(x)$ is $\frac{dy}{dx} = \cos(x)$

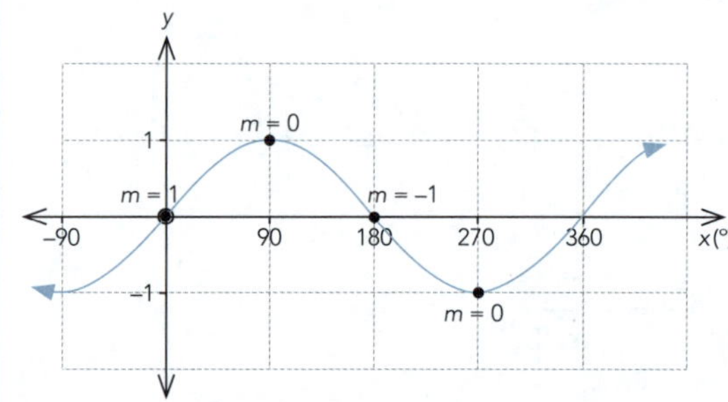

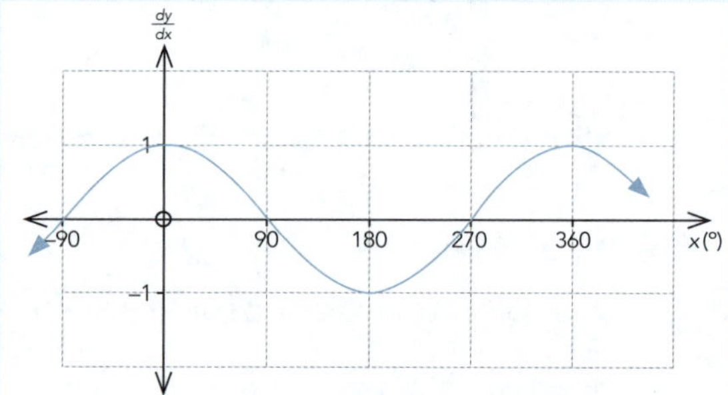

2 The gradient function of $y = \cos(x)$ is $\frac{dy}{dx} = -\sin(x)$

Notice that the sine curve is upside down.

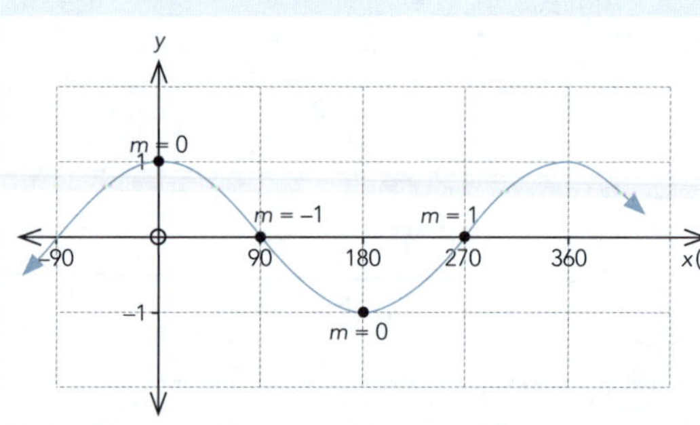

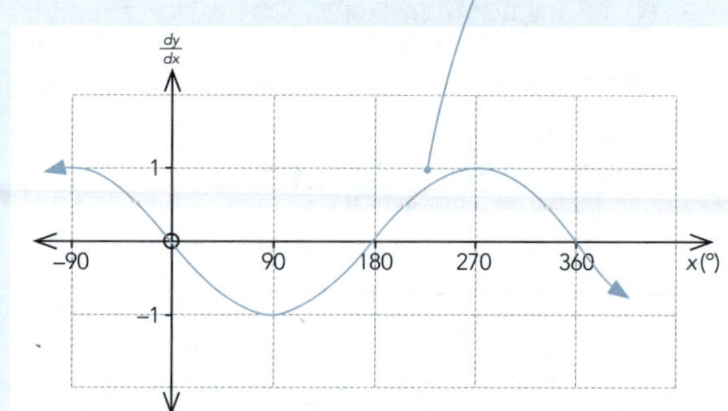

- Using the relationship $\tan(x) = \frac{\sin(x)}{\cos(x)}$ along with the quotient rule, it can be shown that the following are true:

$y = f(x)$	$\frac{dy}{dx} = f'(x)$
$\tan(x)$	$\sec^2(x)$
$\text{cosec}(x)$	$-\text{cosec}(x)\cot(x)$
$\sec(x)$	$\sec(x)\tan(x)$
$\cot(x)$	$-\text{cosec}^2(x)$

These are all on your formula sheet.

- As with exponential and logarithmic functions, almost all examples needing differentiation of trigonometric functions involve use of the **chain rule**, and often the **product rule or quotient rule**.

ISBN: 9780170446976

Examples:

1 Function notation

$$f(x) = 3\cos(2x)$$
$$f = 3\cos(g)$$
$$g = 2x$$
$$f'(g).g' = -3\sin(g) \,.\, 2$$
$$\therefore\ f'(x) = -6\sin(2x)$$

2 Leibniz notation

$$y = \tan(x^2 + 1)$$

Let $y = \tan u$ $\qquad \therefore\ u = x^2 + 1$

Then $\dfrac{dy}{du} = \sec^2(u)$ $\qquad \dfrac{du}{dx} = 2x$

$$\therefore\ \frac{dy}{dx} = \sec^2(u) \,.\, 2x$$
$$= \sec^2(x^2 + 1) \,.\, 2x$$
$$= 2x\sec^2(x^2 + 1)$$

Differentiate the following.

1 $y = \sin(2x)$

2 $f(x) = 4 - \cos(7x)$

3 $y = \cot(5x)$

4 $f(x) = \sec(3x) + 1$

5 $f(x) = \tan(8x - 3)$

6 $f(x) = \mathrm{cosec}(x^2)$

7 $y = 6\cot(2x^3)$

8 $y = \sec(-2x)$

ISBN: 9780170446976

6 The product rule

- This is used where two functions are multiplied together: $f(x).g(x)$

$$(f.g)' = f.g' + g.f' \quad \text{or} \quad \text{if } y = uv, \text{ then } \frac{dy}{dx} = u\frac{dv}{dx} + v\frac{du}{dx}$$

These are on your formula sheet.

Examples:

1 Function notation

$f(x) = x^3(4 - 7x^2) = g.h$

Your formula sheet uses the letters *f* and *g*. Equations in questions are often expressed as $f(x) = \ldots$ These '*f*'s have different meanings ∴ we must use other letters to show our reasoning.

Let $g = x^3$ $\quad h = 4 - 7x^2$

Then $g' = 3x^2$ $\quad h' = -14x$

But $f'(x) = g.h' + h.g'$

so $f'(x) = x^3.(-14x) + (4 - 7x^2).3x^2$

$= -14x^4 + 12x^2 - 21x^4$

$\therefore \; f'(x) = -35x^4 + 12x^2$

2 Leibniz notation

$y = 3x^4 e^{5x} = u \,.\, v$

Let $u = 3x^4$ $\quad v = e^{5x}$

Then $\frac{du}{dx} = 12x^3$ $\quad \frac{dv}{dx} = 5e^{5x}$

But $\frac{dy}{dx} = u\frac{dv}{dx} + v\frac{du}{dx}$

so $\frac{dy}{dx} = 3x^4 \,.\, 5e^{5x} + e^{5x} \,.\, 12x^3$

$\therefore \; \frac{dy}{dx} = 15x^4 e^{5x} + 12x^3 e^{5x}$

3 $f(x) = 10x^2\ln(2x)$

$f'(x) = 10x^2 \,.\, 2 \,.\, \frac{1}{2x} + \ln(2x) \,.\, 20x$

$f'(x) = 10x + 20x\ln(2x)$

4 $y = \tan^2(2x)$

$= \tan(2x) \,.\, \tan(2x)$

Rewrite the square as a product, and use the product rule.

$\frac{dy}{dx} = \tan(2x) \,.\, 2\sec^2(2x) + \tan(2x) \,.\, 2\sec^2(2x)$

$= 4\tan(2x) \,.\, \sec^2(2x)$

 ISBN: 9780170446976

Differentiate the following. Some questions will require the use of the chain rule.

1 $f(x) = x^5(4x - 5)$

2 $f(x) = (3x + 1)(x - 7)$

3 $y = (8x + 5)(3 - 2x)$

4 $y = (x^2 - 6)(7x + 1)$

5 $f(x) = \sqrt{x}\,(9x - 2)$

6 $f(x) = (3x - 5)(x^2 + 4x - 1)$

7 $f(x) = x^3e^x$

8 $f(x) = 5x^4e^{2x}$

9 $y = e^{7x}(x^2 + 4)$

10 $y = e^{2x} \,.\, \sin(x)$

ISBN: 9780170446976

More complex questions which involve use of the product rule and the chain rule:

Examples:

5 Function notation

$f(x) = (3x^2 - 1)^4 \,.\, (2x + 5)^3 = g.h$

Let $\quad g = (3x^2 - 1)^4 \qquad h = (2x + 5)^3$

Then $\quad g' = 4(3x^2 - 1)^3 \,.\, 6x \qquad h' = 3(2x + 5)^2 \,.\, 2$ — Using the chain rule.

$\quad = 24x(3x^2 - 1)^3 \qquad = 6(2x + 5)^2$

But $\quad (g.h)' = g.h' + h.g'$

so $\quad f'(x) = (3x^2 - 1)^4 \,.\, 6(2x + 5)^2 + (2x + 5)^3 \,.\, 24x(3x^2 - 1)^3$

$\quad = 6(3x^2 - 1)^4 (2x + 5)^2 + 24x(2x + 5)^3 (3x^2 - 1)^3$

6 Leibniz notation

$y = \frac{2}{x} \,.\, \sqrt{x^3 - 4} = u.v$

Let $\quad u = \frac{2}{x} \qquad v = \sqrt{x^3 - 4}$

Then $\quad \frac{du}{dx} = \frac{-2}{x^2} \qquad \frac{dv}{dx} = \frac{1}{2}(x^3 - 4)^{-\frac{1}{2}} \,.\, 3x^2$ — Using the chain rule.

$\quad = \frac{3x^2}{2\sqrt{x^3 - 4}}$

But $\quad \frac{dy}{dx} = u \,.\, \frac{dv}{dx} + v \,.\, \frac{du}{dx}$

so $\quad \frac{dy}{dx} = \frac{2}{x} \,.\, \frac{3x^2}{2\sqrt{x^3 - 4}} + \sqrt{x^3 - 4} \,.\, \frac{-2}{x^2}$

$\quad = \frac{3x}{\sqrt{x^3 - 4}} - \frac{2\sqrt{x^3 - 4}}{x^2}$

ISBN: 9780170446976

11 $f(x) = e^{x}(4x^2 - 3x + 1)$

12 $f(x) = e^{2x}(5x - 9)^3$

13 $f(x) = e^{3x^2}(2x - 1)^2$

14 $y = e^{\sqrt{x}}(2x^3 - 5x + 1)$

15 $f(x) = x\ln(2x)$

16 $y = \sin(x)\ln(6x)$

17 $y = x^3\ln(2x - 5)$

18 $f(x) = 3x + 4x\ln(x^2)$

19 $f(x) = 20x - 15x\ln(3x)$

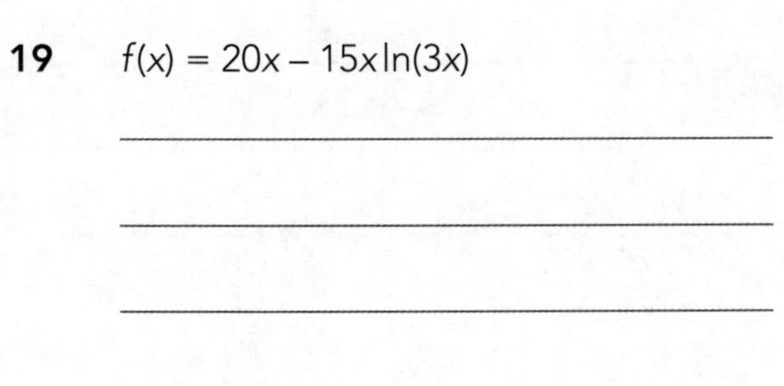

20 $y = e^{x}\ln(5x)$

21 $f(x) = x^2\sin(x)$

22 $y = 2x^3\text{cosec}(3x)$

23 $f(x) = e^x\sec(5x)$

24 $f(x) = (7x + 2)\cos(3x)$

25 $y = e^{4x} \,.\, \tan(5x)$

26 $f(x) = (3x - 1)^2\cos(2x)$

27 $y = (2x - 3)^5(x + 4)^2$

28 $y = (5x - 2)^3(x^2 + 1)$

29 $y = (x + 2)^3 \,.\, \cos(x)$

30 $y = 6xe^{\sqrt{x}}$

ISBN: 9780170446976

7 The quotient rule

- This is used where one function is divided by another: $\dfrac{f(x)}{g(x)}$

$$\left(\frac{f}{g}\right)' = \frac{g.f' - f.g'}{g^2} \quad \text{or} \quad \text{if } y = \frac{u}{v}, \text{ then } \frac{dy}{dx} = \frac{v\dfrac{du}{dx} - u\dfrac{dv}{dx}}{v^2}$$

These are on your formula sheet.

Examples:

1 Function notation

$$f(x) = \frac{3x^2}{2x-5} = \frac{g}{h}$$

Let $g = 3x^2$ $\quad h = 2x - 5$

Then $g' = 6x$ $\quad h' = 2$

But $f'(x) = \dfrac{h.g' - g.h'}{h^2}$

so $f'(x) = \dfrac{(2x-5) \,.\, 6x - 3x^2 \,.\, 2}{(2x-5)^2}$

$= \dfrac{12x^2 - 30x - 6x^2}{(2x-5)^2}$

$= \dfrac{6x^2 - 30x}{(2x-5)^2}$

2 Leibniz notation

$$y = \frac{\sqrt{x} - 4}{6x} = \frac{u}{v}$$

Let $u = \sqrt{x} - 4$ $\quad v = 6x$

Then $\dfrac{du}{dx} = \dfrac{1}{2\sqrt{x}}$ $\quad \dfrac{dv}{dx} = 6$

But $\dfrac{dy}{dx} = \dfrac{v\dfrac{du}{dx} - u\dfrac{dv}{dx}}{v^2}$

Remember: $\dfrac{x}{\sqrt{x}} = \sqrt{x}$

so $\dfrac{dy}{dx} = \dfrac{6x \,.\, \dfrac{1}{2\sqrt{x}} - (\sqrt{x} - 4) \,.\, 6}{36x^2}$

$= \dfrac{3\sqrt{x} - 6\sqrt{x} + 24}{36x^2}$

$= \dfrac{-3\sqrt{x} + 24}{36x^2}$

Divide everything by 3.

$= \dfrac{-\sqrt{x} + 8}{12x^2}$

3 $f(x) = \dfrac{10}{\cos(x)}$

Use the quotient rule.

$$f'(x) = \frac{\cos(x) \,.\, 0 - 10 \,.\, (-\sin(x))}{\cos^2(x)}$$

$$= \frac{10\sin(x)}{\cos(x)} \cdot \frac{1}{\cos(x)}$$

$$= 10\tan(x)\sec(x)$$

or

Use the trigonometric identity to get rid of the fraction.

$$f(x) = \frac{10}{\cos(x)}$$

$$= 10\sec(x)$$

$$f'(x) = 10\sec(x)\tan(x)$$

Differentiate the following.

1 $y = \dfrac{x+5}{x-3}$

2 $f(x) = \dfrac{2x-7}{x+4}$

3 $y = \dfrac{8x}{x-1}$

4 $f(x) = \dfrac{x^2}{5x+2}$

5 $y = \dfrac{x}{\sin(x)}$

6 $y = \dfrac{2x}{\cos(x)}$

7 $f(x) = \dfrac{e^x}{\sin(x)}$

8 $y = \dfrac{2\tan(x)}{x^2}$

 ISBN: 9780170446976

Use both the quotient rule and the chain rule to differentiate the following. You do not need to simplify your answers.

9 $f(x) = \frac{(x-3)^2}{2x}$

10 $y = \frac{8x}{(2x-1)^2}$

11 $y = \frac{e^x}{2x}$

12 $f(x) = \frac{e^{5x}}{x-3}$

13 $y = \frac{x}{e^{3x}}$

14 $y = \frac{e^{x^2}}{4-7x}$

15 $f(x) = \frac{3x^2}{\tan(x)}$

16 $f(x) = \frac{\cos(2x)}{x^3}$

17 $y = \frac{e^{2x}}{\tan(3x)}$

18 $f(x) = \frac{\sin(5x)}{e^{3x-2}}$

ISBN: 9780170446976

8 Parametric functions

- These are functions where **x and y are each defined in terms of a third parameter, t.**
- $\frac{dy}{dx}$ can be found by differentiating each expression and combining the results using the chain rule.

Examples:

1 Let $x = t^3$ $\quad y = 2t$

Then $\frac{dx}{dt} = 3t^2$ $\quad \frac{dy}{dt} = 2$

So $\frac{dy}{dx} = \frac{dy}{dt} \cdot \frac{dt}{dx}$

$\frac{dy}{dx} = 2 \times \frac{1}{3t^2}$

$= \frac{2}{3t^2}$

2 Let $x = \cos(2t)$ $\quad y = \tan^2(t)$

Then $\frac{dx}{dt} = -2\sin(2t)$ $\quad \frac{dy}{dt} = 2\tan(t)\sec^2(t)$

So $\frac{dy}{dx} = \frac{dy}{dt} \cdot \frac{dt}{dx}$

$\frac{dy}{dx} = \frac{2\tan(t)\sec^2(t)}{-2\sin(2t)}$

$= \frac{-\tan(t)}{\sin(2t)\cos^2(t)}$

$\sec^2(t) = \frac{1}{\cos^2(t)}$

You are often required to evaluate this expression. This is easier if it contains only sines, cosines and tangents.

To find the second derivative: $\frac{d^2y}{dx^2}$

- Notice that we **cannot differentiate using the normal method** because $\frac{dy}{dx}$ is expressed in terms of **t**, not **x**.

  However, if we do, the result is $\frac{d\left(\frac{dy}{dx}\right)}{dt}$, or $\frac{d}{dt}\left(\frac{dy}{dx}\right)$.

  Using the chain rule, this can be converted to $\frac{d^2y}{dx^2}$ by multiplying by $\frac{dt}{dx}$:

$$\frac{d}{\cancel{dt}}\left(\frac{dy}{dx}\right)\frac{\cancel{dt}}{dx} = \frac{d^2y}{dx^2}$$

$$\therefore \frac{d^2y}{dx^2} = \frac{d}{dt}\left(\frac{dy}{dx}\right) \cdot \frac{dt}{dx}$$

Example (continued from Example 1 above):

1 $\frac{dy}{dx} = \frac{2}{3t^2} \Rightarrow \frac{d}{dt}\left(\frac{dy}{dx}\right) = \frac{-4}{3t^3}$

$\frac{d^2y}{dx^2} = \frac{-4}{3t^3} \cdot \frac{1}{3t^2}$

$= \frac{-4}{9t^5}$

ISBN: 9780170446976

Find $\frac{dy}{dx}$ and $\frac{d^2y}{dx^2}$ for the following pairs of parametric equations.

1 $x = 5t$ $y = 3t^2$

2 $x = t^2 - 1$ $y = 4t^3$

3 $x = \ln(t)$ $y = 4t^3$

4 $x = \frac{1}{t^2}$ $y = 5t^2 + 4$

Find the first derivative only for the following pairs of parametric equations.

5 $x = e^{2t}$ $y = 3t^4$

6 $x = t^2$ $y = \sqrt{t + 3}$

7 $x = 2t^3$ $y = 3e^{2t}$

8 $x = \ln(2t)$ $y = 2t^3 - t^2$

9 $x = 3\sin(t)$ $y = \cos(2t)$

10 $x = 3\tan(t)$ $y = 2\sin(t)$

11 $x = \sin^2(t)$ $y = 4\cos(t)$

12 $x = 2\sec(t)$ $y = 4\tan(t)$

ISBN: 9780170446976

Applications

Tangents and normals

- The gradient of a curve at a point is the same as the gradient of the tangent at that point.

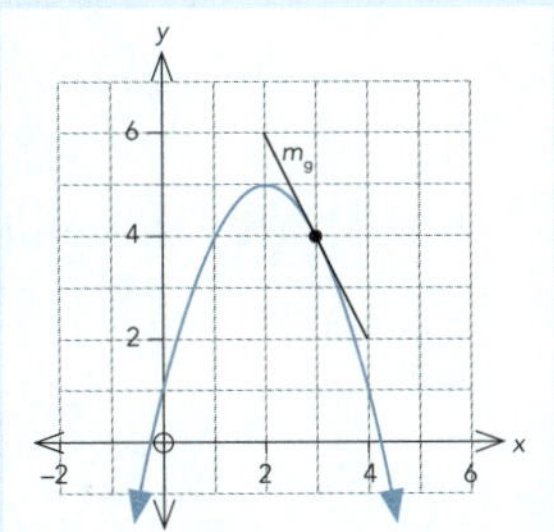

- The normal to a curve at a point is perpendicular to the tangent at that point.

$$\therefore m_g \times m_n = -1$$

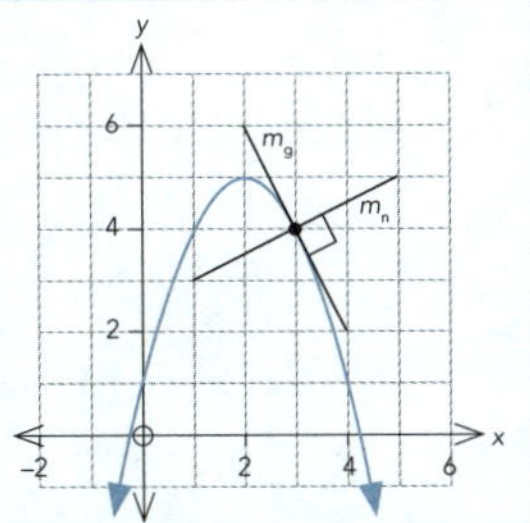

1 Finding gradients and equations of tangents and normals

Examples:

1 Find the equation of the tangent to the curve $y = \sqrt{2x - 1}$ at (5, 3).

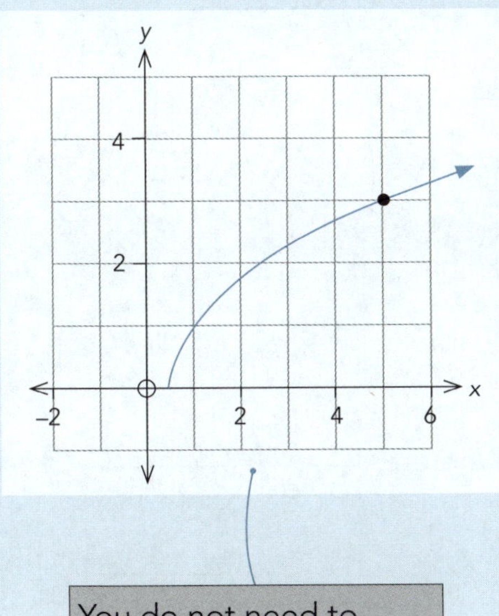

You do not need to draw the graph, but it's useful to check if your answer looks right.

Equation: $y = \sqrt{2x - 1} = (2x - 1)^{\frac{1}{2}}$

Step 1: Differentiate the function to get the gradient function.

$$\frac{dy}{dx} = \frac{1}{2}(2x - 1)^{-\frac{1}{2}} . 2$$

$$= \frac{1}{\sqrt{2x - 1}}$$

In this case you need to use the chain rule.

Step 2: Substitute the value for x.

At (5, 3), $m = \frac{1}{\sqrt{2x - 1}} = \frac{1}{\sqrt{2(5) - 1}}$

$$m = \frac{1}{3}$$

Step 3: Use $y - y_1 = m(x - x_1)$ to find the equation.

$$y - 3 = \frac{1}{3}(x - 5)$$

$$y = \frac{1}{3}x + \frac{4}{3}$$

ISBN: 9780170446976

2 Find the gradient of the normal to the curve $y = \ln(5 - x^2)$ where $x = 1$.

Equation: $y = \ln(5 - x^2)$

Step 1: Differentiate the function to get the gradient function.

$$\frac{dy}{dx} = -2x \cdot \frac{1}{5 - x^2}$$

$$= \frac{-2x}{5 - x^2}$$

Once again, you need to use the chain rule.

Step 2: Substitute the value for x.

At $x = 1$, $m_g = \dfrac{-2x}{5 - x^2}$

$$= \frac{-2(1)}{5 - 1^2}$$

$$m_g = -\frac{1}{2}$$

Step 3: Calculate the gradient of the normal.

$$m_g \times m_n = -1$$

$$-\frac{1}{2} \times m_n = -1$$

$$\therefore \quad m_n = 2$$

3 A curve is defined by $x = 5\sin(t)$ and $y = 3\tan(t)$. Find the gradient of the tangent where $t = \frac{\pi}{3}$.

Step 1: Find an expression for the gradient function.

$$x = 5\sin(t) \Rightarrow \frac{dx}{dt} = 5\cos(t) \quad \text{and} \quad y = 3\tan(t) \Rightarrow \frac{dy}{dt} = 3\sec^2(t)$$

$$\therefore \quad \frac{dy}{dx} = \frac{dy}{dt} \cdot \frac{dt}{dx}$$

$$= \frac{3\sec^2(t)}{5\cos(t)}$$

$$= \frac{3}{5\cos^3(t)}$$

Remember: $\sec^2(t) = \dfrac{1}{\cos^2(t)}$

Step 2: Substitute the value for t.

Where $t = \frac{\pi}{3}$, $\dfrac{dy}{dx} = \dfrac{3}{5\cos^3(t)}$

$$= \frac{3}{5\cos^3\left(\frac{\pi}{3}\right)}$$

$$m = 4.8$$

Remember: $\cos\frac{\pi}{3} = 0.5$

ISBN: 9780170446976

Answer the following questions.

1 A curve has the equation $y = (x^2 - 3x)^3$. Find the gradient of the tangent to the curve at the point where $x = 1$.

2 A curve has the equation $y = (2x - x^2)^3$. Find the equation of the tangent to the curve at the point (1, 1).

3 A curve has the equation $y = x - \frac{4}{x}$.

Find the equation of the normal to the curve at the point (1, – 3).

4 A curve has the equation $y = \frac{2}{x} + \frac{1}{x^2}$.

Find the gradient of the tangent to the curve at the point where $x = 1$.

ISBN: 9780170446976

5 A curve has the equation $y = \frac{e^x}{x+1}$.

Find the gradient of the normal to the curve at the point where $x = 0$.

6 A curve has the equation $y = 4x - e^x$. Find the equation of the tangent to the curve at the point where $x = 0$.

7 A curve has the equation $y = 6\ln(2x - 3)$. Find the gradient of the normal to the curve at the point where $x = 3$.

8 A curve has the equation $y = x\ln x + 1$. Find the equation of the tangent to the curve at the point where $x = 2$.

ISBN: 9780170446976

9 A curve has the equation $y = 2\cos(x)$. Find the gradient of the normal to the curve at the point where $x = \frac{\pi}{6}$.

10 A curve has the equation $y = \sin(2x)$. Find the gradient of the tangent to the curve at the point where $x = \frac{\pi}{3}$.

11 A curve has the equation $y = x\tan(x)$. Find the gradient of the tangent to the curve at the point where $x = \frac{\pi}{4}$.

12 A curve is defined by the parametric equations $x = t^2$ and $y = t^3 + 3$. Find the gradient of the tangent to the curve at a point where $t = 12$.

ISBN: 9780170446976

13 A curve is defined by the parametric equations $x = \frac{1}{(3-t)^2}$ and $y = 5t - t^2$.
Find the gradient of the normal to the curve at a point where $t = 2$.

14 A curve is defined by the parametric equations $x = 3e^{2t}$ and $y = 3e^{4t}$. Find the gradient of the tangent to the curve at a point where $t = 0$.

15 A curve is defined by the parametric equations $x = \sqrt{t+4}$ and $y = \sin\frac{t}{2}$.
Find the gradient of the tangent to the curve at a point where $t = 0$.

16 A curve is defined by the parametric equations $x = \tan^2(t)$ and $y = \sin(2t)$. Find the gradient of the tangent to the curve at a point where $t = 0$.

ISBN: 9780170446976

2 Finding coordinates of points given the equation and gradient

Examples:

1 Find the x coordinate of the point on the curve $f(x) = e^{2x} - 10x$ where the tangent to the curve is parallel with the x-axis.

Equation: $f(x) = e^{2x} - 10x$

Step 1: Differentiate the function to get the gradient function. $\frac{dy}{dx} = 2e^{2x} - 10$

Step 2: Tangent is parallel to x-axis $\Rightarrow \frac{dy}{dx} = 0$

$$\begin{aligned} \therefore\ 2e^{2x} - 10 &= 0 \\ 2e^{2x} &= 10 \\ e^{2x} &= 5 \\ \log_e e^{2x} &= \log_e 5 \\ 2x &= \ln 5 \\ x &= 0.8047 \end{aligned}$$

2 The graph below shows the function $y = \sqrt{20 - x^2}$ where $y \geq 0$ and the tangent to the function at point (2, 4). Find the coordinates of P, the point where the tangent crosses the x-axis.

Equation: $y = \sqrt{20 - x^2} = (20 - x^2)^{\frac{1}{2}}$

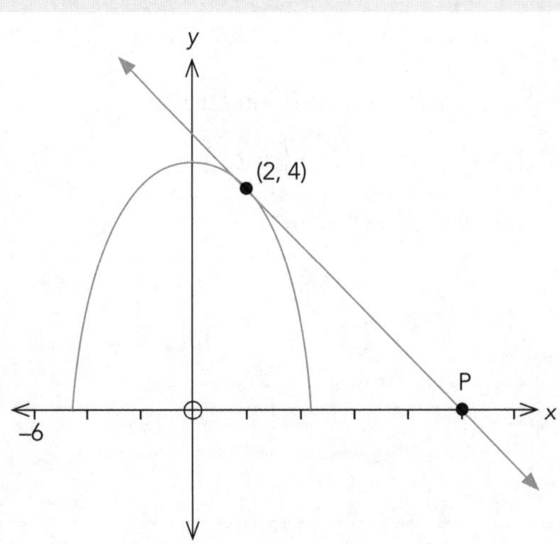

Step 1: Differentiate the function to get the gradient function.

$$\begin{aligned} \frac{dy}{dx} &= \frac{1}{2} \cdot \frac{1}{\sqrt{20 - x^2}} \cdot -2x \\ &= \frac{-x}{\sqrt{20 - x^2}} \end{aligned}$$

Step 2: Find the equation of the tangent using the point (2, 4) and the gradient.

$$\begin{aligned} m &= \frac{-2}{\sqrt{20 - 2^2}} \\ &= -\frac{1}{2} \end{aligned}$$

$$\therefore\ y - 4 = -\frac{1}{2}(x - 2)$$

$$y = -\frac{1}{2}x + 5$$

Step 3: Find the coordinates of P.

$$y = 0 \Rightarrow 0 = -\frac{1}{2}x + 5$$

$$x = 10$$

$\therefore$ Coordinates of P are (10, 0).

ISBN: 9780170446976

Answer the following questions.

1 Find the x coordinate of the point on the curve $f(x) = 3e^{4x} - 24x$ where the tangent to the curve is parallel to the x-axis.

2 Find the x coordinate at which the tangent to the function $y = \dfrac{3}{e^{2x-1}} + 6x$ has a gradient of 0.

3 A curve is defined by the parametric equations $x = 2t^2 + 3$ and $y = \dfrac{t^3}{3} - 9t$. Find the coordinates of the point(s) on the curve for which the tangent has a gradient of 0.

4 For what x value(s) is the tangent to the graph of the function $y = \dfrac{x-2}{x(x+6)}$ parallel to the x-axis?

 ISBN: 9780170446976

5 The tangent to the curve $y = \sqrt{x - 2}$ is drawn at the point (6, 2). Find the coordinates of the point P where the tangent intersects the x-axis.

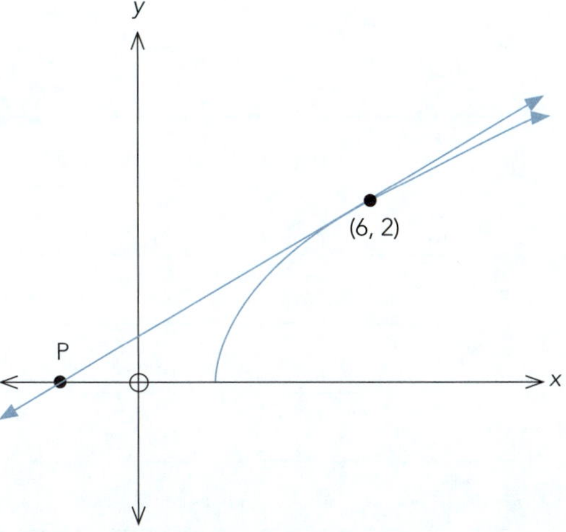

6 For $0 < y < \frac{\pi}{2}$, find the coordinates of the point on the curve $y = \cos(x)$ where the normal to the curve has a gradient of $\frac{2}{\sqrt{3}}$.

7 Two tangents to the curve $y = \frac{1}{2}(x^2 - 2x + 3)$ pass through the point (5, 7). Find the coordinates of the points where these tangents touch the curve.

8 The normal to the parabola $y = -0.5(x+4)^2 + 3$ at the point (– 2, 1) intersects the parabola again at point P. Find the x coordinate of P.

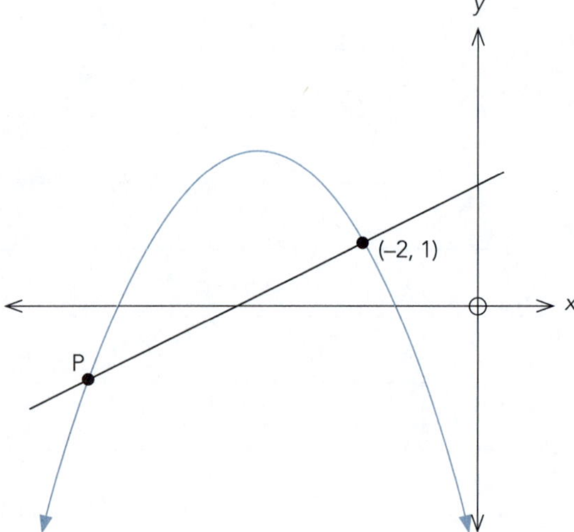

9 The tangents to the curve $y = \frac{(x-4)^2}{4} + 3$ at the points P and Q are perpendicular. P is the point (8, 7). Find the x coordinate of Q.

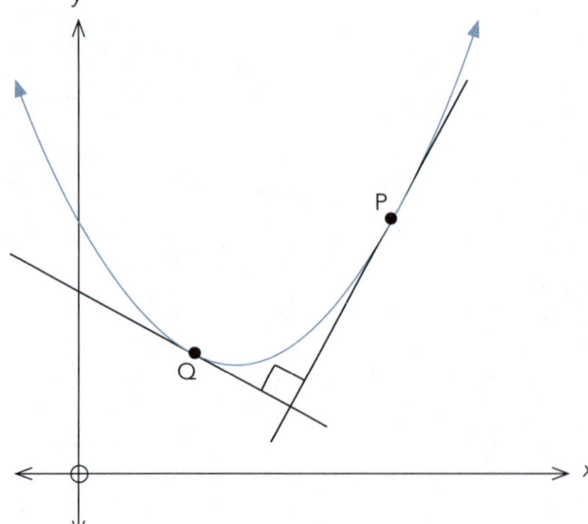

ISBN: 9780170446976

Stationary points

1 Finding stationary points

- These occur where the **gradient** of the function **equals zero**.

There are three types of stationary points:

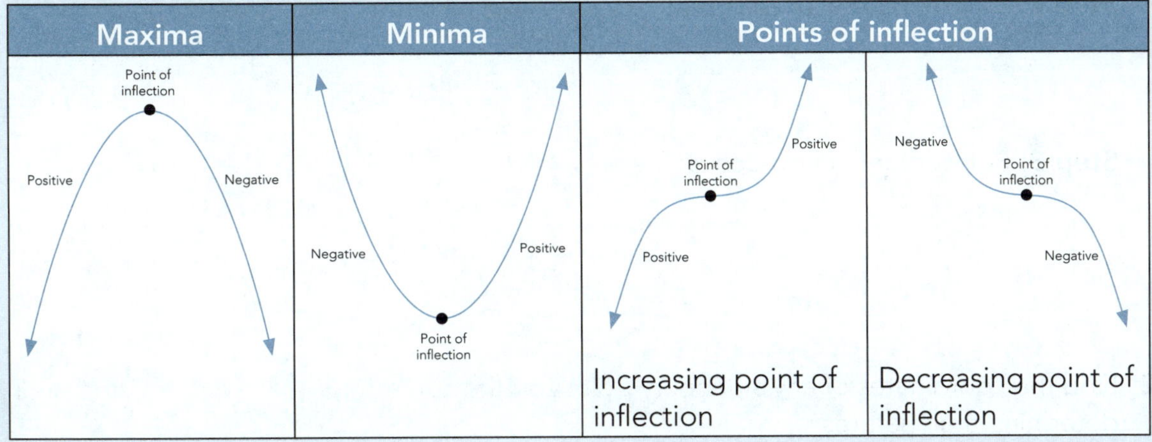

At all of these points, the gradient = 0, so $f'(x) = \frac{dy}{dx} = 0$.

Using second derivatives to determine the nature of stationary points

- We consider the **change** in the gradient, which is found by **differentiating a second time**.
- The second derivative is written as $f''(x)$ or $\frac{d^2y}{dx^2}$.

1 **Maxima**
The gradient is **decreasing** (positive to negative) $\Rightarrow f''(x) = \frac{d^2y}{dx^2}$ is **negative**.

2 **Minima**
The gradient is **increasing** (negative to positive) $\Rightarrow f''(x) = \frac{d^2y}{dx^2}$ is **positive**.

3 **Points of inflection**
These may be increasing, decreasing or stationary: $f''(x) = \frac{d^2y}{dx^2}$ is **0**.

> To remember these, recall quadratic graphs: $y = +x^2 \ldots \Rightarrow$ minimum $y = -x^2 \ldots \Rightarrow$ maximum

Terminology: A **local** maximum (or minimum) occurs where the turning point is not the highest (or lowest) point on the curve.

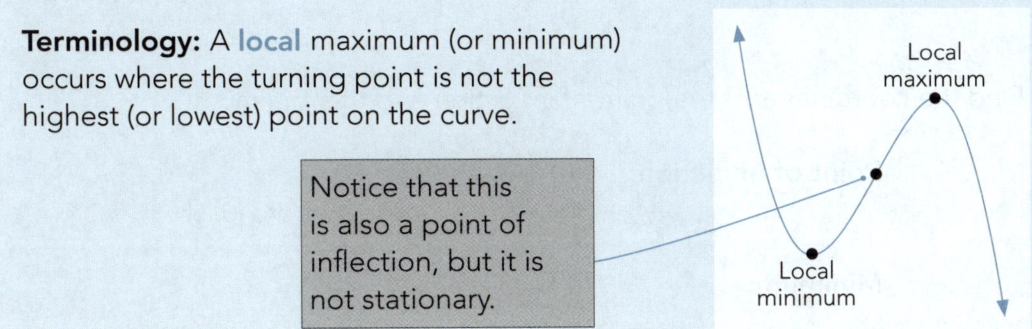

Examples:

1 Find the *x* values for any point(s) of inflection on the graph of the function $y = e^{-x^2-1}$.

Equation: $y = e^{-x^2-1}$

Step 1: Differentiate the function to get the gradient function. $\dfrac{dy}{dx} = -2xe^{x^2-1}$

Step 2: Differentiate a second time: $\dfrac{d^2y}{dx^2} = -2x \,.\, -2xe^{x^2-1} + e^{x^2-1} \,.\, -2$

$$= 2e^{x^2-1}(2x^2 - 1)$$

Step 3: Points of inflection occur where $\dfrac{d^2y}{dx^2} = 0$ $\quad \therefore\ 2e^{x^2-1}(2x^2 - 1) = 0$

$$2x^2 = 1$$

$$x = \pm\sqrt{\frac{1}{2}}$$

Remember:
e^{x^2-1} cannot $= 0$

2 Find the coordinates of any stationary points on the curve $f(x) = 3x^4 - 8x^3 + 5$, and determine their natures.

Equation:

Step 1: Differentiate the function to get the gradient function: $f(x) = 3x^4 - 8x^3 + 5$

$$\therefore\ f'(x) = 12x^2(x - 2)$$

Step 2: Find the *x* coordinates of any stationary points: $f'(x) = 12x^2(x - 2) = 0$

$$\therefore\ x = 0 \text{ or } x = 2$$

Step 3: Determine their nature: $f''(x) = 36x^2 - 48x$

$$= 12x(3x - 4)$$

Where $x = 0$: $\quad f''(0) = 12(0)[3(0) - 4]$

$= 0 \Rightarrow$ Point of inflection

Where $x = 2$: $\quad f''(2) = 12(2)[3(2) - 4]$

$= +48 \Rightarrow$ Minimum

You do not need to calculate the value (48): you just need to show that it is positive.

Step 4: Find the coordinates of the point of inflection and the minimum:

Point of inflection: $f(0) = 3(0)^4 - 8(0)^3 + 5$

$= 5 \ \therefore$ Coordinates are (0, 5)

Minimum: $f(2) = 3(2)^4 - 8(2)^3 + 5$

$= -11 \ \therefore$ Coordinates are (2, –11)

ISBN: 9780170446976

Answer the following questions.

1 **a** Find the coordinates of any stationary point(s) on the curve $f(x) = 3x^4 - 4x^3 + 1$.

b Determine the nature of the stationary point(s).

2 **a** If $f(x) = 9\ln(x - 2.5) - 20x$, find the x coordinate for the turning point where $0 \leq x \leq 5$.

b Determine the nature of this turning point, and find its coordinates.

ISBN: 9780170446976

3 **a** For what value of k does the function $y = x + e^x + \frac{k}{x}$ have a stationary point where $x = 1$?

b Determine the nature of this turning point, and find its coordinates.

4 For the function $f(x) = \frac{25}{x-3} + x$, find the x value(s) of any stationary points.

5 Find the x coordinates of the stationary points of the function $y = 2x + \frac{8}{x^2}$.

ISBN: 9780170446976

2 Optimisation

- The optimum value for a function always lies at a **maximum** or **minimum** point.

Steps:

1 If you aren't given a diagram, then draw one, if it is possible.
2 Identify what needs to optimised, and write an equation for it.
3 Find equations that link any other variables in this equation.
4 Write the equation in terms of **one** variable only (otherwise it won't be differentiable).
5 Differentiate to find the optimum and, if required, determine its nature.
6 **Answer in the context of the question.**

Examples:

1 One side of a rectangle lies on the x-axis, and the vertices of its opposite side lie above the x-axis and on the curve $y = 6 - x^2$. Calculate the maximum possible area of the rectangle.

Step 1: Draw a picture:

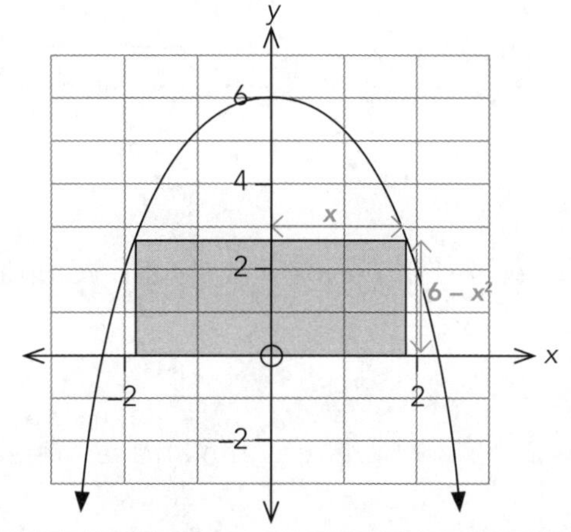

Step 2: Identify what needs to optimised, and write an equation for it:

Area needs to be optimised. $\text{Area} = 2xy$

Step 3: Equation linking variables: $y = 6 - x^2$

Step 4: Write equation in terms of one variable:

$$\text{Area} = 2x(6 - x^2) = 12x - 2x^3$$

Step 5: Differentiate:

$$\frac{dA}{dx} = 12 - 6x^2 = 0$$

$$\therefore \quad x^2 = 2$$

$$x = \pm\sqrt{2}$$

Step 6: Answer in the context of the question:

$$x = \sqrt{2}$$

$$\therefore \quad \text{Maximum area} = 12\sqrt{2} - 2(\sqrt{2})^3$$

$$= 12\sqrt{2} - 4\sqrt{2}$$

$$= 8\sqrt{2} \text{ square units}$$

2 An open box is to be formed by cutting squares (with sides x cm) from the corners of a rectangular piece of card, and folding the sides up. The card measures 25 cm by 40 cm.

- Find the values of x that result in boxes with maximum or minimum volumes.
- Show which value(s) result in maximum volumes.
- Find the maximum volume of the box.

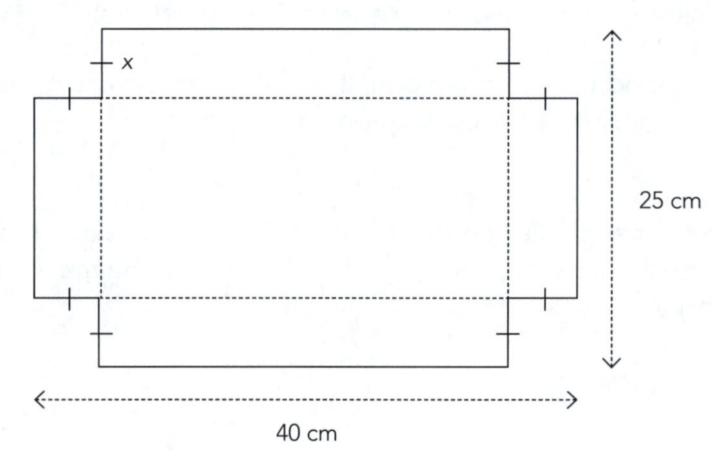

Step 1: Find the equation for the volume:

$$V = x(40 - 2x)(25 - 2x)$$
$$= 1000x - 130x^2 + 4x^3$$

Step 2: Find where maxima and minima occur:

$$\frac{dV}{dx} = 12x^2 - 260x + 1000$$
$$= (x - 5)(3x - 50)$$
$$\therefore \quad x = 5 \text{ or } x = 16.\dot{6}$$

Step 3: Determine the nature of these points:

$$\frac{d^2V}{dx^2} = 24x - 260$$

Where $x = 5$:

$$\frac{d^2V}{dx^2} = 24(5) - 260$$
$$= -140 \Rightarrow \text{Maximum}$$

Where $x = 16.\dot{6}$:

$$\frac{d^2V}{dx^2} = 24(16.\dot{6}) - 260$$
$$= +140 \Rightarrow \text{Minimum}$$

Step 4: Find the maximum volume:

$$V = x(40 - 2x)(25 - 2x)$$
$$= 5(40 - 10)(25 - 10)$$
$$= 2250 \text{ cm}^3$$

ISBN: 9780170446976

3 Find the maximum volume for a cone that has a slant height of 10 cm.

Step 1: Draw a picture:

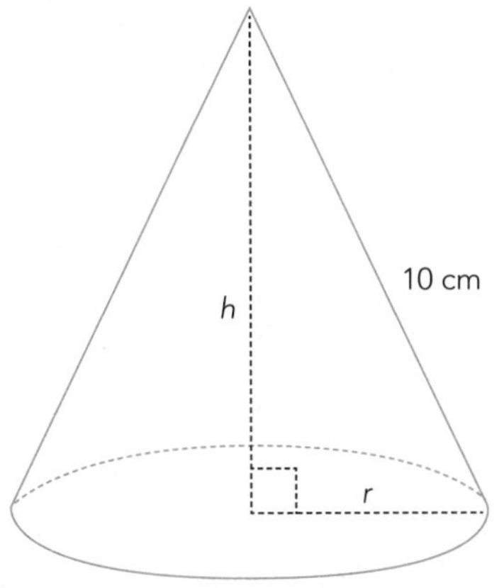

Step 2: Identify what needs to optimised, and write an equation for it:

Volume needs to be optimised. Volume $= \frac{1}{3}\pi r^2 h$

This is on your formula sheet.

Step 3: Equation linking variables: $r^2 + h^2 = 100$

Step 4: Write equation in terms of one variable:

We could use $r = \sqrt{100 - h^2}$ or $h = \sqrt{100 - r^2}$

Often you have a choice of variables. It pays to think carefully about which will make calculations easier.

Substituting for r will be easier because there is an r^2 in the formula, so the square root will disappear:

$$V = \frac{1}{3}\pi(100 - h^2)h$$
$$= \frac{1}{3}\pi(100h - h^3)$$

Step 5: Differentiate:

$$\frac{dV}{dh} = \frac{1}{3}\pi(100 - 3h^2) = 0$$
$$\therefore\ 3h^2 = 100$$
$$h = \sqrt{\frac{100}{3}}$$
$$= 5.774 \text{ cm}$$
$$r = \sqrt{100 - \frac{100}{3}}$$
$$= 8.165 \text{ cm}$$

Step 6: Answer in the context of the question:

$$\text{Maximum volume} = \frac{1}{3}\pi\,(8.165)^2\,(5.774)$$
$$= 403.1 \text{ cm}^3$$

Answer the following questions.

1 The decline in the population of a disease-ravaged colony of insects was modelled by the equation $P = 2x + \frac{50}{x}$, where x represents the number of days since the outbreak, and $1 \leq x \leq 6$. When was the population at its minimum, and how many insects remained?

2 A right-angled triangle has one vertex at the origin, one on the curve $y = (x - 6)^2$, and the third at a point on the x-axis where $0 < x < 6$. Find the maximum possible area for the triangle, and the coordinates of the vertex that lies on the curve.

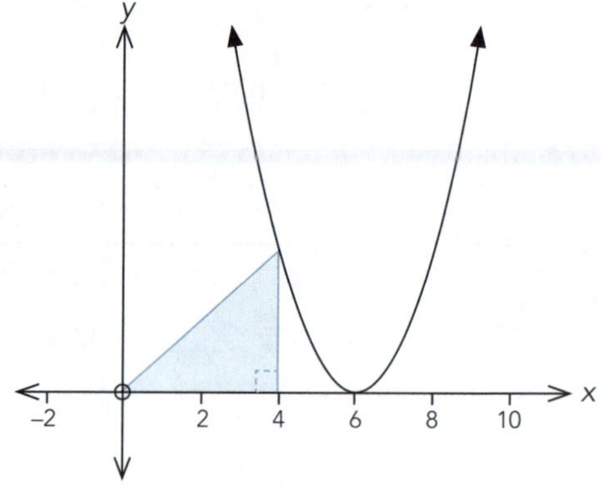

ISBN: 9780170446976

3 The rate of growth of a tree depends on its age. For a particular species in Nelson, it can be modelled by the equation $G = 40 - 0.5x - \frac{72}{x}$, $4 \le x \le 20$.

G represents the rate of growth in cm per year and x represents its age in years. Find the turning point for this model, determine its nature and explain its meaning in terms of the growth of this tree.

4 Find the coordinates of the point $P(x, y)$ on the curve $y = 2\sqrt{x}$ that is closest to the point (6, 0).

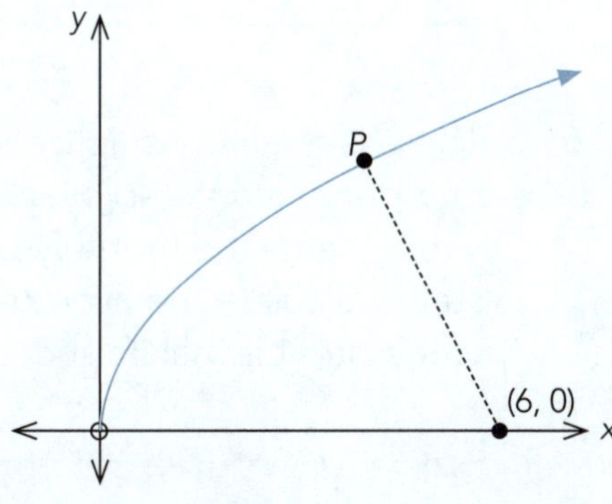

5 An isosceles triangle is created between the origin and points on the curve $y = 12 - x^2$. Find the maximum possible area of the triangle, and coordinates of A and B.

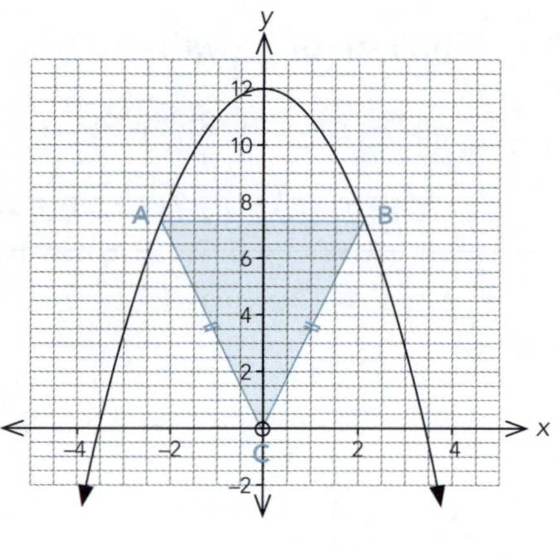

6 There is a relationship between the monthly profit made by a café and the average number of staff employed. These are related by the equation $y = 15x - 6x\ln(x)$, where y represents the monthly profit (x \$1000), and x represents the average number of staff. Calculate the average number of staff that should be employed so that the café's profit is maximised.

ISBN: 9780170446976

7 A shape is formed from a rectangle and a semicircle. Its total perimeter is 200 m. Find the values for r and x that maximise its area, where x represents the length of the rectangle and r represents the radius of the semicircle.

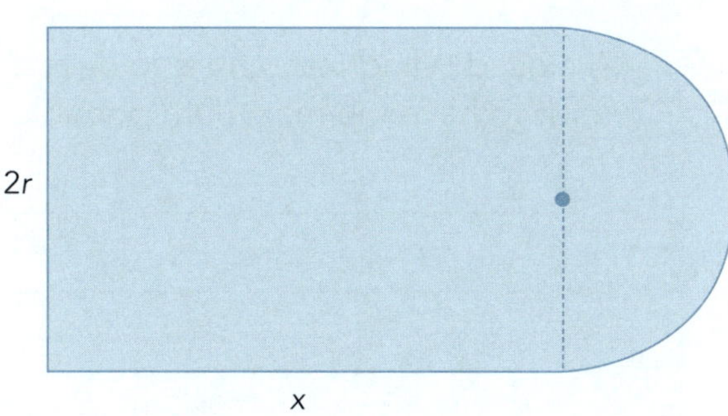

8 A rectangle is drawn with vertices that lie on the circle whose equation is $x^2 + y^2 = 36$. Find the maximum possible area of the rectangle, and coordinates of point B.

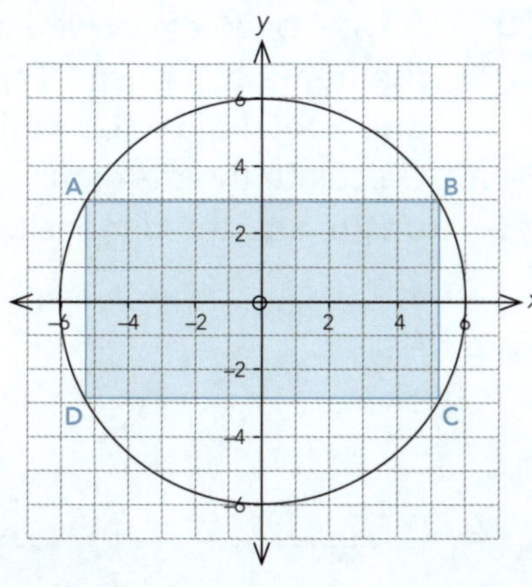

ISBN: 9780170446976

9 A closed cylindrical tank is to have a surface area of 24 m^2. Find the radius of the tank for it to have the maximum possible volume.

10 A total of 36 cm of wire is to be used to form the frame of a prism. The ends of the prism are equilateral triangles with edges that are x cm long. Calculate the maximum possible volume of the prism, and the dimensions at which this occurs.

x

 ISBN: 9780170446976

Related rates

- Related rates occur when the **change in one parameter results in changes in other parameters**.
  Examples: If the radius of a sphere changes, then its surface area and its volume also change.
- Rates of change are at the heart of calculus.
  Examples: $\frac{dr}{dt}$ means the change in the radius with respect to time.
  $\frac{dV}{dr}$ means the change in the volume with respect to the radius.
- If you have two rates, you can use the **chain rule** to find a third rate:
  Example: $\frac{dV}{dt} = \frac{dr}{dt} \cdot \frac{dV}{dr}$, which represents the change in volume with respect to time.

1 Questions involving two rates

Example: The volume of a bubble is increasing at 2 cm³/s. At what rate is the radius increasing when its radius is 3 cm?

Given: 'volume of a bubble is increasing at 2 cm³/s' $\Rightarrow \frac{dV}{dt} = 2$ cm³/s

'**when** its radius is 3 cm' $\Rightarrow r = 3$

Do not use the '**when**' piece of information until the **end**.

Asked to find: 'At what rate is the radius increasing' $\Rightarrow$ find $\frac{dr}{dt}$.

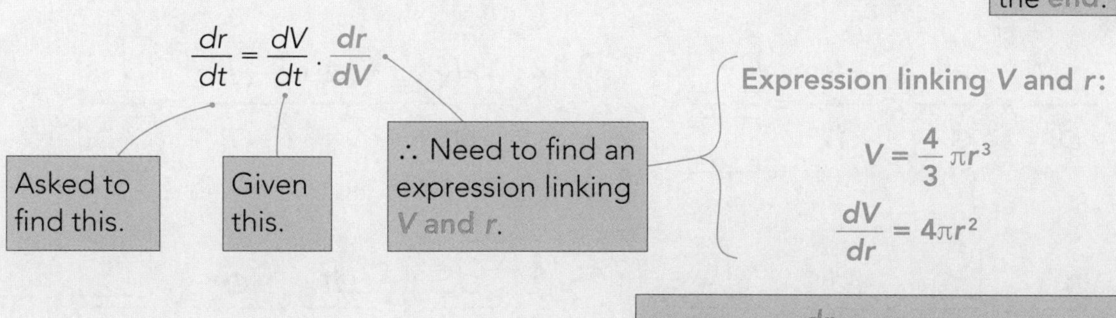

$$\frac{dr}{dt} = 2 \cdot \frac{1}{4\pi r^2} = \frac{1}{2\pi r^2}$$

Notice that $\frac{dr}{dV}$ was needed, so the expression has to be inverted: $\frac{dr}{dV} = \frac{1}{\frac{dV}{dr}} = \frac{1}{4\pi r^2}$

When $r = 3$, $\frac{dr}{dt} = \frac{1}{18\pi} = 0.01768$ cm/s

Remember, do not substitute the value from the '**when**' requirement until **after** you have an expression for the required rate.

A Fill in the boxes in the following equations.

1 $\dfrac{dV}{dr} = \dfrac{dt}{dr} \times \dfrac{dV}{\square}$

2 $\dfrac{dr}{dt} = \dfrac{dV}{\square} \times \dfrac{dr}{dV}$

3 $\dfrac{dA}{dt} = \dfrac{\square}{dt} \times \dfrac{dA}{dx}$

4 $\dfrac{\square}{dt} = \dfrac{dx}{dt} \times \dfrac{dP}{dx}$

5 $\dfrac{\square}{\square} = \dfrac{dV}{dr} \times \dfrac{dr}{dx}$

6 $\dfrac{\square}{\square} = \dfrac{dx}{dt} \times \dfrac{dP}{dx}$

7 $\dfrac{dA}{dt} = \dfrac{\square}{\square} \times \dfrac{dA}{dx}$

8 $\dfrac{dS}{da} = \dfrac{\square}{\square} \times \dfrac{dx}{da}$

9 $\dfrac{dV}{da} = \dfrac{dt}{\square} \times \dfrac{\square}{dt}$

10 $\dfrac{dA}{dr} = \dfrac{dx}{dr} \times \dfrac{\square}{\square}$

11 $\dfrac{dV}{dr} \times \dfrac{dt}{dx} \times \dfrac{dr}{dt} = \dfrac{\square}{\square}$

12 $\dfrac{dx}{dr} \times \dfrac{dA}{dx} \times \dfrac{dr}{dt} = \dfrac{\square}{\square}$

13 $\dfrac{dA}{dt} = \dfrac{\square}{dt} \times \dfrac{dr}{dV} \times \dfrac{dA}{\square}$

14 $\dfrac{dS}{dr} = \dfrac{dt}{dr} \times \dfrac{dx}{\square} \times \dfrac{\square}{dx}$

15 $\dfrac{dV}{dt} = \dfrac{dr}{dx} \times \dfrac{dV}{dr} \times \dfrac{\square}{\square}$

16 $\dfrac{dA}{dr} = \dfrac{dA}{ds} \times \dfrac{dt}{dr} \times \dfrac{\square}{\square}$

17 $\dfrac{dP}{dr} = \dfrac{dt}{dx} \times \dfrac{dx}{dr} \times \dfrac{\square}{\square}$

18 $\dfrac{dS}{da} = \dfrac{dr}{dt} \times \dfrac{dS}{dr} \times \dfrac{\square}{\square}$

ISBN: 9780170446976

B Answer the following questions.

1 The sides of a square are increasing at a rate of 5 mm/s. How fast is the area increasing when the side of the square is 12 mm?

2 The sides of a cube are shrinking at 0.1 cm/min. Calculate the rate of change in its volume when the sides are 6 cm long. (Hint: 'shrinking' $\Rightarrow$ the change is negative.)

3 A stone is dropped into a pond. The radius of the resulting circular ripple increases at a rate of 24 cm/s. Find how fast the area is changing when the radius of the ripple is 10 cm.

ISBN: 9780170446976

4 A sphere is being inflated at a constant rate of 8 cm^3/s. Find the rate at which the radius is increasing when the radius is 5 cm.

5 A cube initially has sides of 1 cm. It expands at a rate of 3 cm/min. Calculate the rate of change in its volume when its surface area is 47.04 cm^2.

6 A sphere is being inflated at a constant rate of 20 cm^3/s. At what rate is the radius of the sphere increasing when the volume of the sphere is 2304π cm^3?

ISBN: 9780170446976

2 Questions involving ratios

- These questions may involve the use of **trigonometry** or **similar triangles**.

Examples:

1 The shaded triangle (PQX) is drawn within the right-angled triangle PRS, and angle SPR = 60°.

Point Q moves along the base of the triangle PR, beginning at point P, at a constant speed of 2 cm/s.

At what rate is the area of the shaded triangle changing **when** point Q is 10 cm from point P?

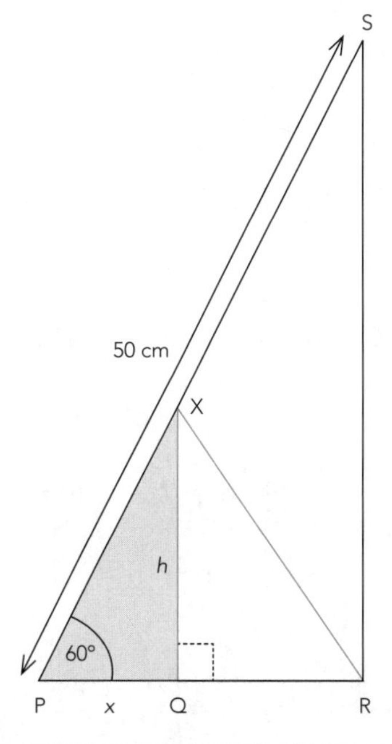

Given: 'at a constant speed of 2 cm/s' $\Rightarrow \frac{dx}{dt} = 2$ cm/s

'**when** point Q is 10 cm from point P' $\Rightarrow x = 10$

Asked to find: 'At what rate is the area of the shaded triangle changing' $\Rightarrow$ find $\frac{dA}{dt}$.

$$\frac{dA}{dt} = \frac{dx}{dt} \cdot \frac{dA}{dx}$$

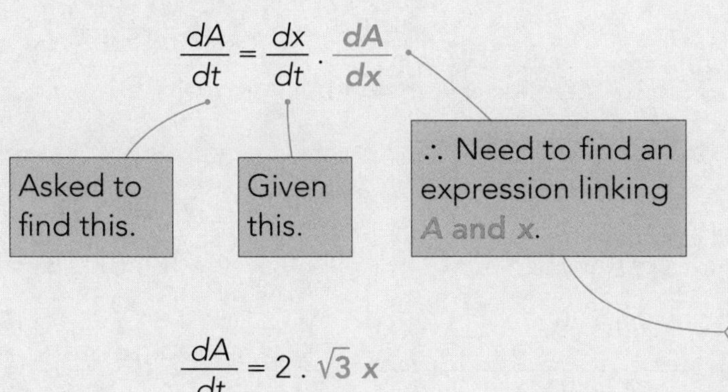

Expression linking A and x:

$$A = \frac{1}{2} . b . h$$
$$= \frac{1}{2} . x . h$$

XPQ = 60° $\Rightarrow \tan 60° = \frac{h}{x}$

$$h = \sqrt{3}\, x$$

$$A = \frac{1}{2} . x . \sqrt{3}\, x$$
$$= \frac{\sqrt{3}}{2} x^2$$
$$\frac{dA}{dx} = \sqrt{3}\, x$$

$$\frac{dA}{dt} = 2 . \sqrt{3}\, x$$

When $x = 10$,

$$\frac{dA}{dt} = 2 . \sqrt{3}\, x$$
$$= 2\sqrt{3} . 10$$
$$= 36.64 \text{ cm}^2/\text{s}$$

$\therefore$ Rate of change in area = 36.64 cm²/s

ISBN: 9780170446976

2 A right circular cone is being filled with liquid at a rate of 85 cm^3 per minute. The angle between the sides of the cone is 84°. Find the rate at which the depth is increasing **when** the liquid is 6 cm deep.

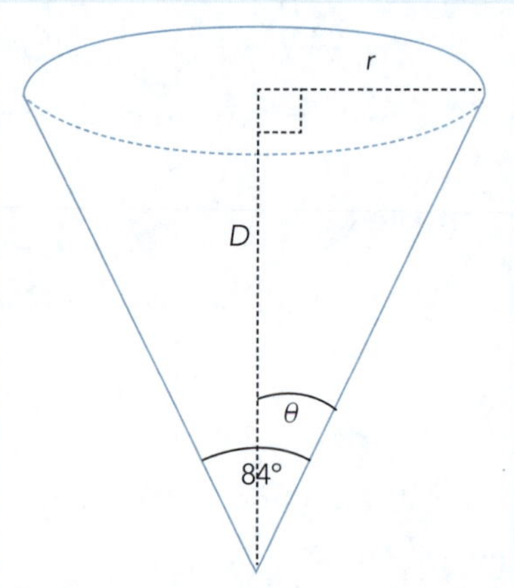

Given: 'is being filled with liquid at a rate of 85 cm^3 per minute' $\Rightarrow \dfrac{dV}{dt} = 85$ cm^3/min

'The angle between the sides of the cone is 84°' $\Rightarrow \theta = 42°$.

'**when** the liquid is 6 cm deep' $\Rightarrow D = 6$

Asked to find: 'the rate at which the depth is increasing' $\Rightarrow$ find $\dfrac{dD}{dt}$.

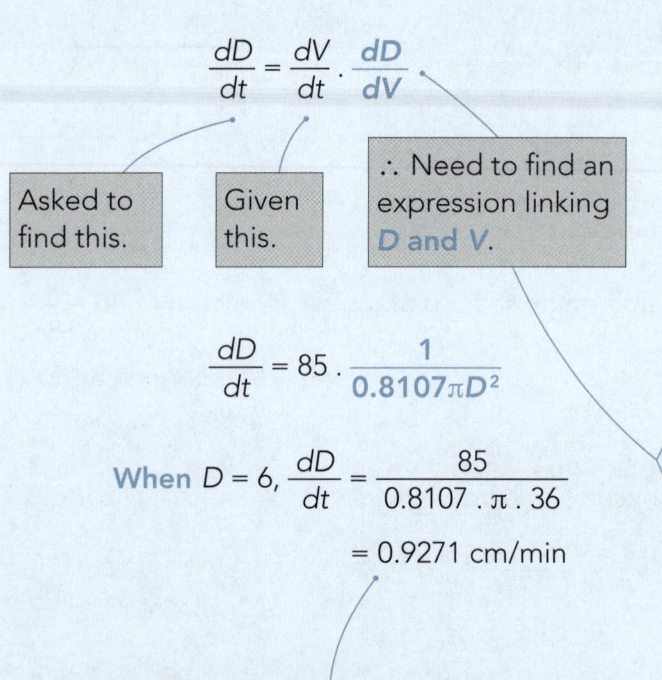

$$\frac{dD}{dt} = \frac{dV}{dt} \cdot \frac{dD}{dV}$$

$$\frac{dD}{dt} = 85 \cdot \frac{1}{0.8107\pi D^2}$$

When $D = 6$, $\dfrac{dD}{dt} = \dfrac{85}{0.8107 \cdot \pi \cdot 36}$

$= 0.9271$ cm/min

Expression linking V and D:

$$V = \frac{1}{3}\pi r^2 D$$

Notice that r cannot be considered a constant because if D changes, so does r ∴ need to express r in terms of D.

$$\tan 42° = \frac{r}{D}$$

$$r = D \cdot \tan 42°$$

$$= 0.9004\,D$$

$$V = \frac{1}{3}\pi r^2 D$$

$$= \frac{1}{3}\pi \cdot (0.9004D)^2 \cdot D$$

$$= \frac{1}{3}\pi \cdot 0.8107 \cdot D^3$$

$$\frac{dV}{dD} = \pi \cdot 0.8107\,D^2$$

$$= 0.8107\,\pi D^2$$

ISBN: 9780170446976

Answer the following questions.

1 Olive and Fred are arborists. Fred is standing 25 m away and watching Olive as she descends from a tree at a rate of 0.8 m per second. θ is the angle of elevation from Fred's eyes to Olive. Find the rate of change of θ when Olive is 25 m above Fred's eye level.

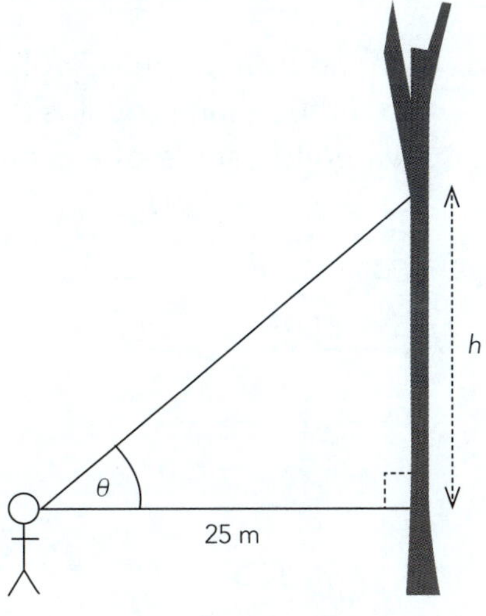

2 A sprinkler that sends out a jet of water is located at point A. The jet of water rotates in an anticlockwise direction at a rate of $\frac{1}{10}$ radians/s. The shortest distance from the sprinkler to a wall (BC) is 8 m. Find the rate at which the jet of water is moving along the wall when $\theta = \frac{\pi}{3}$.

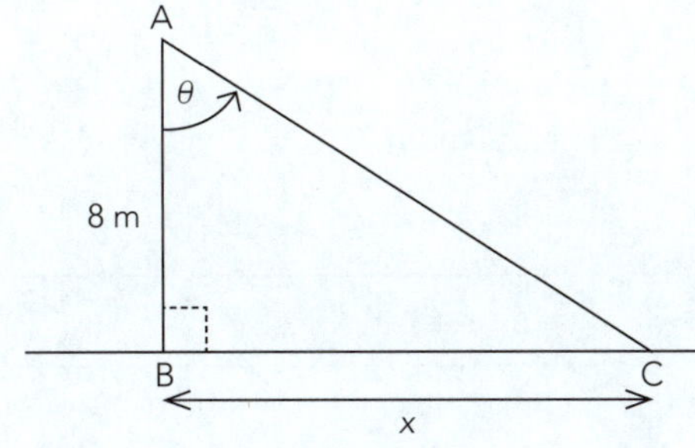

ISBN: 9780170446976

3 In the evening, the angle of elevation of the sun is decreasing at 0.02 radians per minute. A lamp post is 6 m high. Find the rate of change of the lamp post's shadow when the angle of elevation of the sun is $\frac{\pi}{6}$.

4 A 5 m ladder is propped against a wall. The top of the ladder starts sliding down the wall at 0.2 m/s. How fast is the bottom of the ladder moving away from the base of the wall when the top of the ladder is 4 m above the ground?

5 m

h

x

 ISBN: 9780170446976

5 A car is driving towards B at a constant speed of 6 m/s. A camera is located at C, and is focused on the car. If it is to remain focused on the car as it travels towards B, calculate the rate of rotation of the camera when the car is 30 m from A.

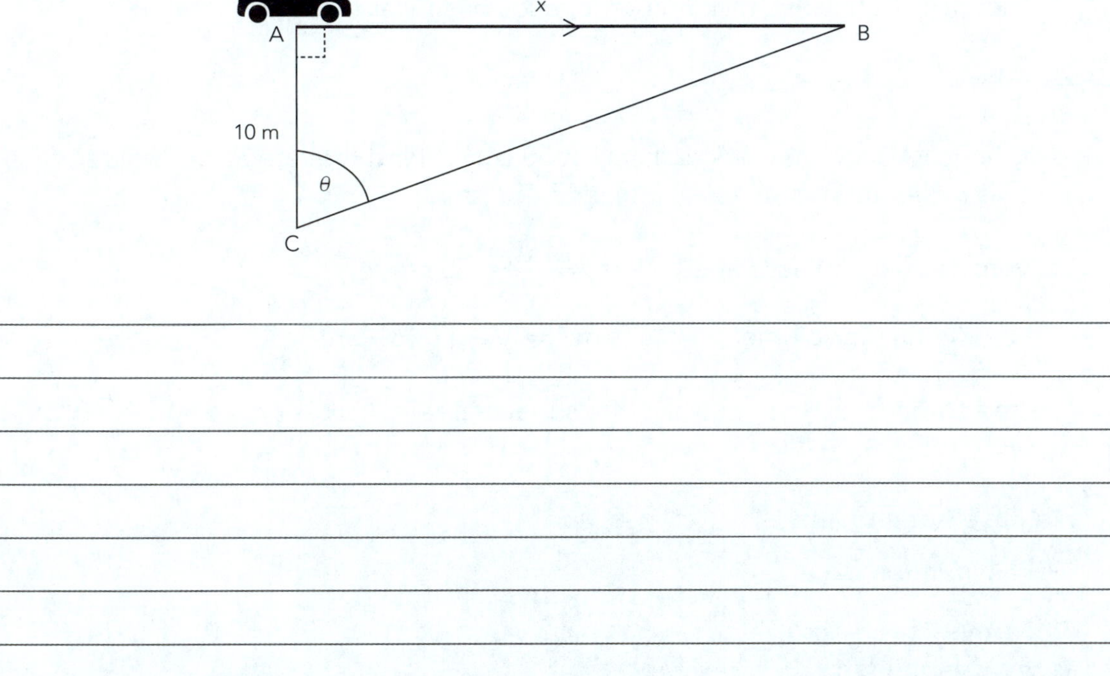

6 A grandfather is teaching his granddaughter to row. He has tied a long rope to the dinghy. He stands on a wharf and lets the rope out as she rows away from him at 0.6 m/s. He lets the rope out from 3 m above the point of attachment to the dinghy.

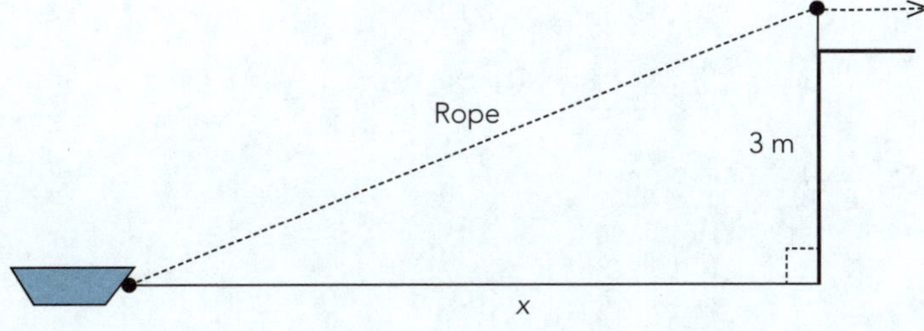

Calculate the rate at which the grandfather is letting the rope out when the granddaughter is 4 m from the wharf.

3 Questions involving three rates

- In these questions, you will need to **multiply three rates in order** to find the answer.
- Some may also involve trigonometric ratios or similar triangles.

Examples:

1 A spherical balloon is being inflated at 50 cm³/s. Find the rate at which its surface area is increasing **when** its volume is 17 481 cm³.

Given: 'is being inflated at 50 cm³/s' $\Rightarrow \dfrac{dV}{dt} = 50$ cm³/s

'**when** its volume is 17 481 cm³' $\Rightarrow V = 17\,481$ cm³

Asked to find: 'the rate at which the surface area is increasing' $\Rightarrow$ find $\dfrac{dA}{dt}$.

We have been given $\dfrac{dV}{dt} \Rightarrow$ we must use a derivative that has dV as the denominator.

We need $\dfrac{dA}{dt} \Rightarrow$ must use a derivative that has dA as the numerator.

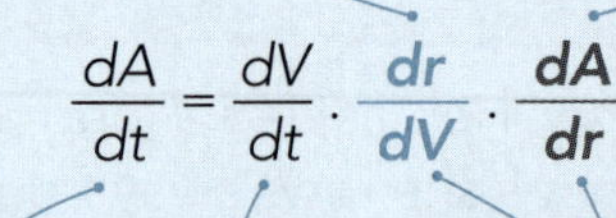

$$\frac{dA}{dt} = \frac{dV}{dt} \cdot \frac{dr}{dV} \cdot \frac{dA}{dr}$$

Asked to find this.

Given this.

$\therefore$ Need to find an expression linking V and r and A and r.

r is the common parameter in the formulas for A and V.

Expression linking V and r:

$$V = \frac{4}{3}\pi r^3$$

$$\frac{dV}{dr} = 4\pi r^2$$

So, $$\frac{dr}{dV} = \frac{1}{4\pi r^2}$$

Expression linking A and r:

$$A = 4\pi r^2$$

$$\frac{dA}{dr} = 8\pi r$$

$$\frac{dA}{dt} = \frac{dV}{dt} \cdot \frac{dr}{dV} \cdot \frac{dA}{dr}$$

$$= 50 \cdot \frac{1}{4\pi r^2} \cdot 8\pi r$$

$$= \frac{100}{r}$$

When $V = 17\,481$ cm³ $\Rightarrow \dfrac{4}{3}\pi r^3 = 17\,481$

$$r = \sqrt[3]{\frac{(3 \times 17\,481)}{4\pi}}$$

$$r = 16.1$$

$$\frac{dA}{dt} = \frac{100}{16.1}$$

$$= 6.211 \text{ cm}^2/\text{s}$$

ISBN: 9780170446976

2 A right circular cone is 27 cm high and it has a radius of 18 cm. It is being filled with liquid at a rate of 48 cm^3/s. At what rate will the surface area of the liquid be increasing when the depth of the liquid is 24 cm?

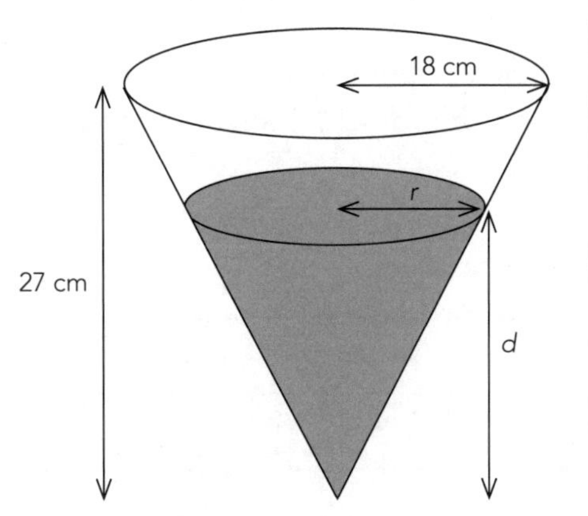

Given: 'is being filled with liquid at a rate of 48 cm^3/s' $\Rightarrow \frac{dV}{dt} = 48$ cm^3/s

'when the depth of liquid is 24 cm' $\Rightarrow d = 24$ cm

Asked to find: 'the rate at which its surface area is increasing' $\Rightarrow$ find $\frac{dA}{dt}$.

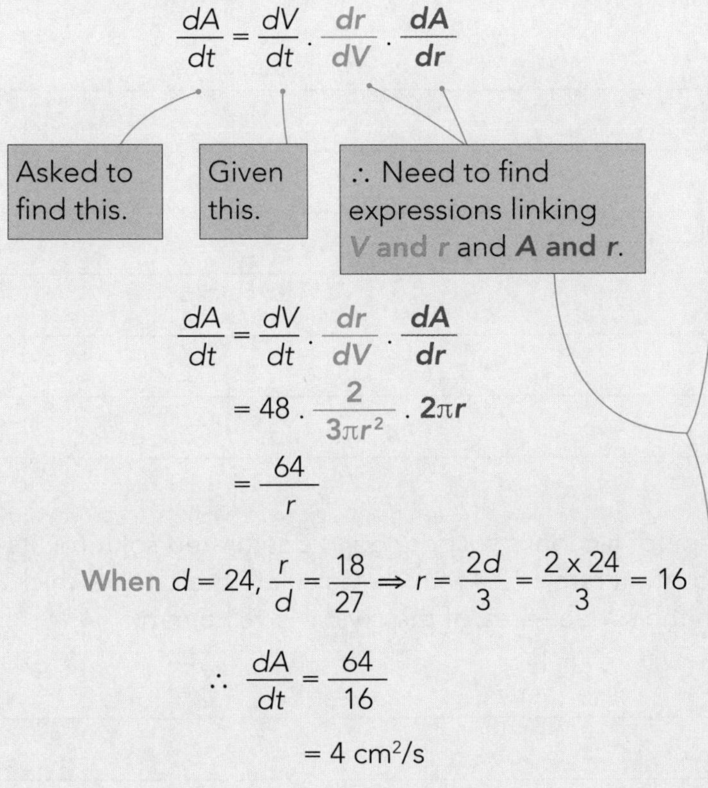

$$\frac{dA}{dt} = \frac{dV}{dt} \cdot \frac{dr}{dV} \cdot \frac{dA}{dr}$$

$$\frac{dA}{dt} = \frac{dV}{dt} \cdot \frac{dr}{dV} \cdot \frac{dA}{dr}$$

$$= 48 \cdot \frac{2}{3\pi r^2} \cdot 2\pi r$$

$$= \frac{64}{r}$$

When $d = 24$, $\frac{r}{d} = \frac{18}{27} \Rightarrow r = \frac{2d}{3} = \frac{2 \times 24}{3} = 16$

$$\therefore \frac{dA}{dt} = \frac{64}{16}$$

$$= 4 \text{ cm}^2/\text{s}$$

***r* is the common parameter in the formulas for *A* and *V*.**

$$\frac{r}{d} = \frac{18}{27} \Rightarrow d = \frac{3r}{2}$$

Expression linking *V* and *r*:

$$V = \frac{1}{3}\pi r^2 d$$

$$= \frac{1}{3}\pi r^2 \cdot \frac{3r}{2}$$

$$= \frac{1}{2}\pi r^3$$

$$\frac{dV}{dr} = \frac{3\pi r^2}{2}$$

So, $\frac{dr}{dV} = \frac{2}{3\pi r^2}$

Expression linking *A* and *r*:

$$A = \pi r^2$$

$$\frac{dA}{dr} = 2\pi r$$

Answer the following questions.

1 The circumference of a circular ripple is growing at a rate of 10 cm per second. Find the rate at which its area is increasing when its radius is 20 cm.

2 The surface area of a spherical bubble is expanding at 5 cm^2/s. Find the rate at which its volume is increasing when its area is 353 cm^2.

3 A crystal that grows in the shape of a cube is suspended in a saturated solution. Its surface area is growing at a constant rate of 3.4 mm^2/h. Calculate the rate at which the volume is increasing when the surface area of the crystal is 181.5 mm^2.

ISBN: 9780170446976

4 Grain is being tipped from the end of a conveyor belt onto a conical pile. The slant of the cone forms an angle of $\frac{\pi}{6}$ radians with the horizontal. The conveyor belt adds grain to the pile at a rate of 4 m^3 per minute. Find the rate at which the area of the base of the pile is increasing when the radius is 5 m.

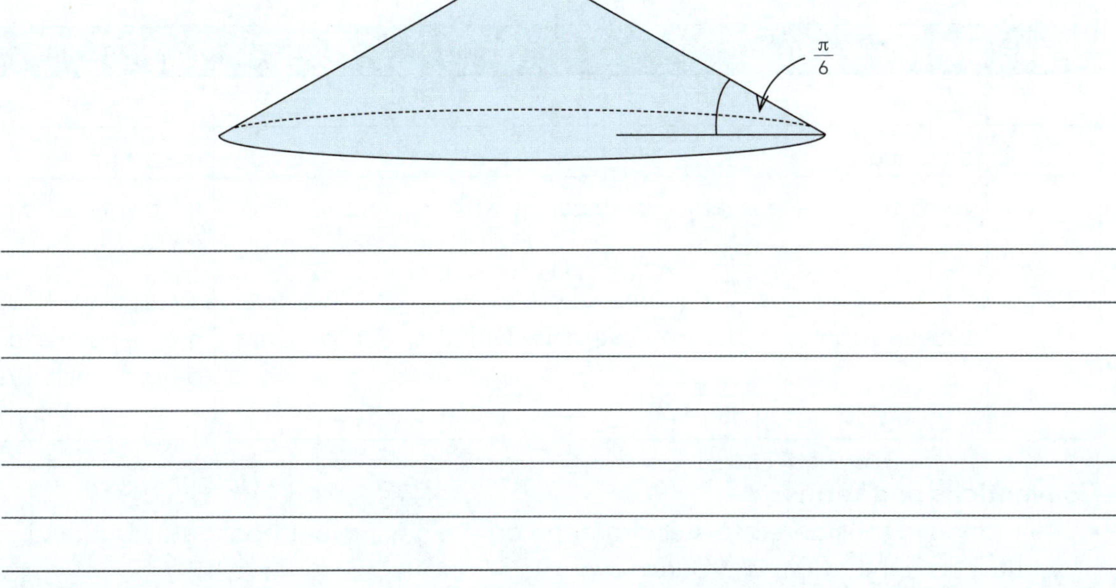

5 A sacrificial anode whose shape is a cuboid is bolted to a boat's hull. Its width is twice its height and its length is three times its height.

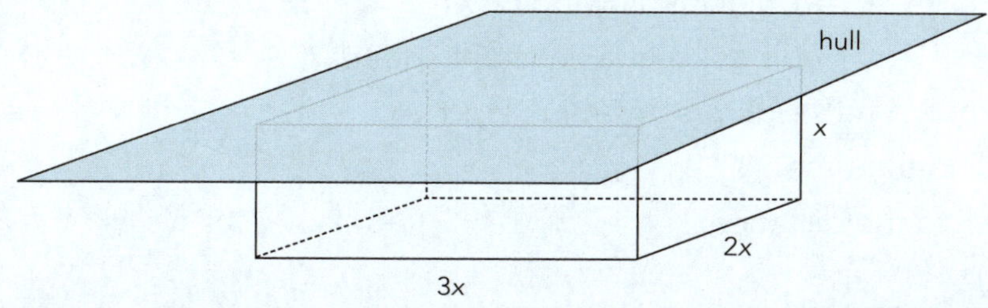

Its surface area decreases at a rate of 2 mm^2 per week. Assuming that it retains its cuboidal shape, calculate the rate of change in its volume when $x = 5$ mm.

ISBN: 9780170446976

Kinematics

Kinematics is the study of motion.

Parameters and units

Letter	What it stands for		Unit
t	time		second (s)
s	distance		metre (m)
v	velocity	= change in distance with respect to time $= \dfrac{ds}{dt} = v(t) = s'(t)$	metre/second (m/s)
a	acceleration	= change in velocity with respect to time $= \dfrac{dv}{dt} = a(t) = v'(t) = \dfrac{d^2s}{dt^2}$	metre per second per second (m/s^2)

Conventions and terms

- All motion is considered to start from the origin. It is usually horizontal, but if it is vertical then the origin is ground level.
- Motion usually starts where time = 0. 'Initially' $\Rightarrow t = 0$.
- Velocity = 0 if the object is 'at rest', 'at its greatest height', or 'stopped'.
  Units: distance/time, e.g. m/s, m/min, km/h
- Acceleration: +ve ⇒ object is speeding up
  –ve ⇒ object is slowing down
  = 0 ⇒ object is at a constant speed
  Units: distance/time^2, e.g. m/s^2, m/min^2, km/h^2
- 'When' ⇒ find the time.
- 'Where' ⇒ find the distance.

Because these interrelationships involve change, we can apply our calculus skills:

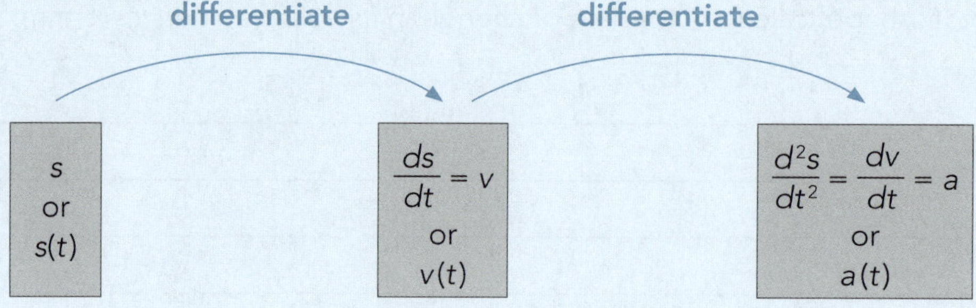

- All questions asked in this standard will require that you differentiate.

ISBN: 9780170446976

Example: The velocity of an object is modelled by the function $v(t) = -\sqrt{t}\,(t^2 - 5t)$, where $v(t)$ is the velocity of the object in m/s and t is the time in seconds since the start of the object's motion. Show that the object's maximum velocity occurs when $t = 3$.

Maximum velocity $\Rightarrow$ $a(t) = 0$

$$a(t) = -\left[\sqrt{t}\,.\,(2t - 5) + (t^2 - 5t)\,.\,\frac{1}{2\sqrt{t}}\right]$$

Differentiate using the product and chain rules.

If $t = 3$, then $a(t) = -\left[\sqrt{3}\,(6 - 5) + (9 - 15)\,.\,\frac{1}{2\sqrt{3}}\right]$

$$= -\left[\sqrt{3} - \frac{6}{2\sqrt{3}}\right]$$

$$= 0$$

$\therefore$ Where $t = 3$, there is a maximum, minimum or point of inflection.

To show that the velocity is a maximum:

Note that differentiating a second time to show that there is a maximum at $t = 3$ would be messy, so substitute values either side of $t = 3$.

$$v(2) = -\sqrt{2}\,(2^2 - 5(2)) = 6\sqrt{2} = 8.49$$

$$v(3) = -\sqrt{3}\,(3^2 - 5(3)) = 6\sqrt{3} = 10.39$$

$$v(4) = -\sqrt{4}\,(4^2 - 5(4)) = 4\sqrt{4} = 8$$

$\therefore$ The object's maximum velocity occurs where $t = 3$.

Answer the following questions.

1 The velocity of an object is modelled by the function $v(t) = (t - 3)^2(2t - 1)$, where $v(t)$ is the velocity of the object in m/s and t is the time in seconds since the start of the object's motion. Find the times when the acceleration of the object is 0.

ISBN: 9780170446976

2 The motion of an object is modelled by the function $s(t) = \frac{2 + t^2}{t}$, where $s(t)$ is the distance of the object from a point in metres and t is the time in seconds since the start of the object's motion. Find when the object is closest to the point.

3 A particle is moving in a straight line. The distance, in metres, travelled by the particle can be modelled by the following function: $s(t) = \ln(t^3 + 7t)$, where $t \geq 0$ and t is measured in seconds. Find the velocity of this particle after 2 seconds.

ISBN: 9780170446976

4 The velocity of an object is modelled by the function $v = \ln\left(t - \frac{t^2}{4}\right) + 6$, where v is the velocity of the object in m/s and t is the time in seconds since the start of the object's motion. Show that after 2 seconds, the object's acceleration is 0.

5 The velocity of an object is modelled by the function $v = 2e^{t} + 5e^{-t}$, where v is the velocity of the object in m/s and t is the time in seconds since the start of the object's motion. Find the time when the acceleration of the object is 0.

Practice questions

Practice question 1

a Differentiate $y = \dfrac{4}{\tan \theta}$

b A curve is defined parametrically by the parametric equations $x = \cos t$ and $y = \dfrac{1}{t^2}$

Find an expression for the gradient of the tangent to this curve in terms of t.

c Find the coordinates of any turning points on the curve $y = 5 - 2x.\ln(x)$, and determine their nature(s).

ISBN: 9780170446976

d A crane is lifting a block of concrete at a constant rate of 1.5 m/s. The supervisor is standing 20 m away, watching the ascent of the block. Let the angle of elevation of the base of the block from the supervisor's eye level be θ.

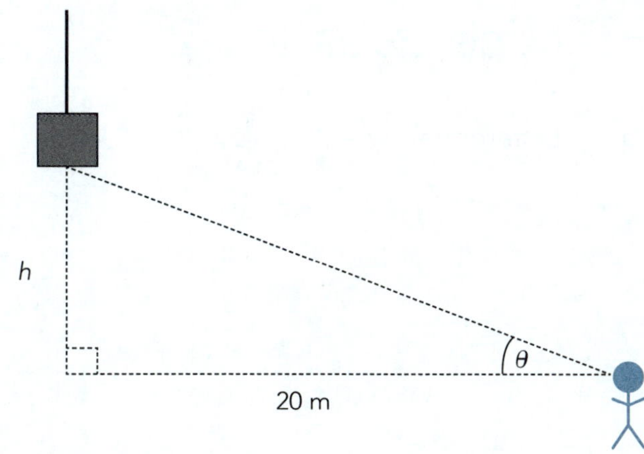

Find the rate at which the angle of elevation is increasing when the base of the block is 10 m above the supervisor's eye level.

e A curve is defined by the function $y = e^{-(x-2)^2}$. Find the x coordinates for any points of inflection on this curve.

ISBN: 9780170446976

Practice question 2

a Differentiate $y = \frac{2}{x^3} - \frac{1}{\sqrt{x}}$

b Find the rate of change of the function $f(t) = 9\ln(5t - t^2)$ when $t = 2$.

c The diagram shows the function $y = f(x)$.

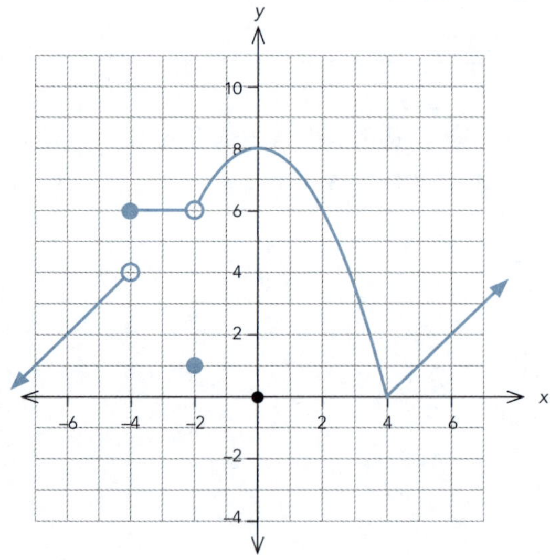

i What is the value of $f(-4)$?

State clearly if the value does not exist.

ii For what value(s) does the function $f(x)$ not have a limit?

iii Find all the value(s), if any, which meet the following conditions:

1 $f(x)$ is continuous but not differentiable.

2 $f'(x) > 0$.

iv What is the value of $\lim_{x \to -2} f(x)$?

State clearly if the limit does not exist.

ISBN: 9780170446976

d The velocity of an object is modelled by the function $v = 3t + 5e^{-2t}$ for $t \geq 0$, where v is the velocity of the object in m/s and t is the time in seconds since the start of the object's motion. Find the time when the acceleration of the object is 0.

e Find the maximum volume of a cone if the sum of its height and its radius is 6 cm.

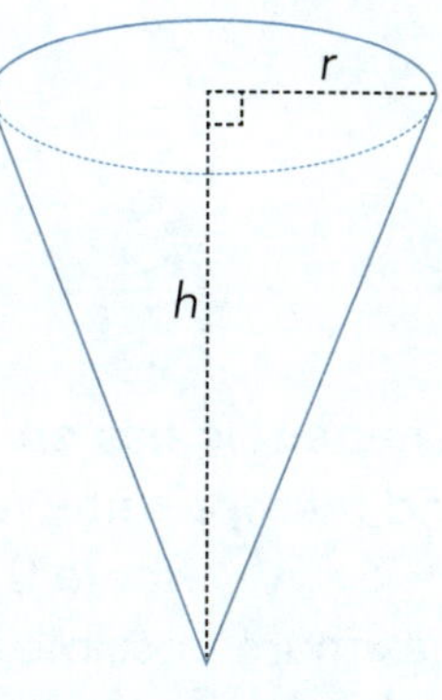

ISBN: 9780170446976

Practice question 3

a Differentiate $f(x) = e^{5x}(3x^2 - 7)$

b Find an expression for the gradient of the normal to the function $f(x) = \frac{8}{\sqrt{x}}$

c A rectangle has one vertex at the origin, and the opposite vertex on the curve $y = 6 - \sqrt{x}$, where $0 \le x \le 36$. Find the maximum possible area of the rectangle, and the coordinates of the vertex that lies on the curve.

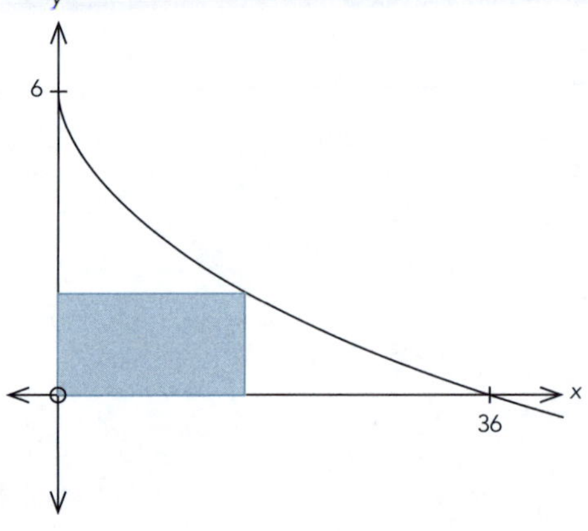

 ISBN: 9780170446976

d If $y = \frac{e^{2x}}{\cos x}$, show that $\frac{dy}{dx} = y(2 + \tan x)$

e The volume of a spherical bubble is increasing at a rate of 12 mm^3 per second. Find the rate at which its surface area is increasing when its radius is 9 mm.

ISBN: 9780170446976

Answers

Revision (pp. 7–14)

Differentiation of polynomials (pp. 8–9)

1 $\frac{dy}{dx} = \frac{2}{3}$

2 $f'(x) = \frac{3x}{4} + 5$

3 $\frac{dy}{dx} = 14x - 4x^5$

4 $f'(x) = x + \frac{1}{8}$

5 $\frac{dy}{dx} = 35 - 40x$

6 $f'(x) = -60x - 27$

7 $\frac{dy}{dx} = -\frac{3}{8}x^2$

8 $f'(x) = 0.06x + 0.08$

9 $\frac{dy}{dx} = -0.6$

10 $f'(x) = \frac{15}{2}x^2 - 10x$

11 $\frac{dy}{dx} = -x^4 + \frac{3}{5}$

12 $f'(x) = 12x^2 - 56x + 49$

Logarithms (pp. 10–14)

Fundamentals (pp. 10–11)

1 $\log_2 16 = 4$

2 $\log_{10} a = 6$

3 $\log_3 9 = b$

4 $\log_5 5 = 1$

5 $\log_c d = 3$

6 $\log_a c = b$

7 $2^5 = a$

8 $b^4 = c$

9 $6^d = c$

10 $d^b = 125$

11 $x^z = y$

12 $c^b = a$

13 $\log_9 1 = 0$

14 $\log_7 7 = 1$

15 $\log_{10} 0.0001 = -4$

16 $\log_{25} 5 = \frac{1}{2}$

17 $\log_6 \frac{1}{6} = -1$

18 $\log_{16} 2 = \frac{1}{4}$

19 $\log_{25} \frac{1}{5} = -\frac{1}{2}$

20 $\log_{81} 9 = 0.5$

21 $5^0 = 1$

22 $5^{-1} = \frac{1}{5}$

23 $49^{\frac{1}{2}} = 7$

24 $10^1 = 10$

25 $64^{\frac{1}{6}} = 2$

26 $10^{-2} = 0.01$

27 $10^{-4} = \frac{1}{10\,000}$

28 $2^{-3} = \frac{1}{8}$

29

Log	Value
$\log_5 200$	3.292
$\log_{100} 14$	0.573
$\log_{10} 200\,000$	5.301
$\log_2 80$	6.322
$\log_7 60$	2.104
$\log_3 90$	4.096
$\log_{23} 30$	1.085

Rules (pp. 12–13)

1 log *a* + log *b* = log *ab* (p. 12)

1 log 35

2 log 60

3 log 30

4 log 10

5 0

6 log *pqr*

7 log *p* + log *q*

8 log *a* + log *b* + log *c*

2 log *a* – log *b* = log $\frac{a}{b}$ (p. 12)

9 $b^d = c$

10 log 15

11 log 10

12 log $\frac{fg}{h}$

13 log 24 – log *b*

14 log *c* – log 2

15 log 27 – log 2 – log *d*

16 log *n* + log *p* – log *q*

3 *n*log *a* = loga^n (p. 13)

17 log 81

18 log 9

19 log 3

20 log $b^{\frac{1}{a}}$

Putting the rules together (pp. 13–14)

21 4

22 $\frac{9}{2}$ log 2 or $\frac{3}{2}$ log 8

23 $\frac{1}{5}$

24 log 5

25 $\frac{7}{3}$

26 0

27 $\frac{1}{3}$

28 $\frac{2}{3}$

29 –11log 10

30 0

Concepts (pp. 16–27)

Sketching gradient functions (pp. 17–19)

Note: Dots represent points that must be correct.

1 a

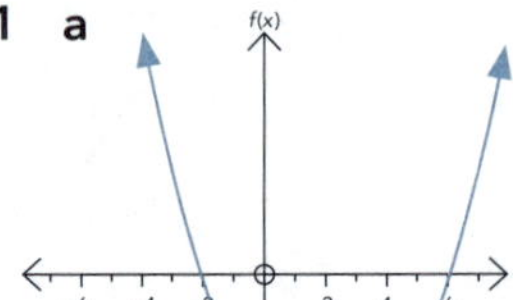

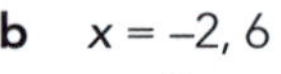

b $x = -2, 6$

c $x = 2$

d No values

 ISBN: 9780170446976

2 a

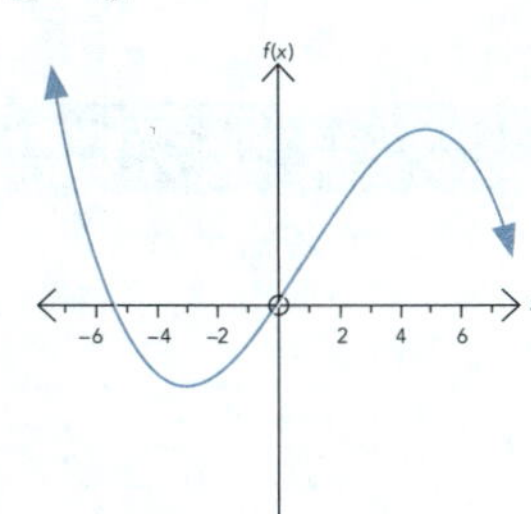
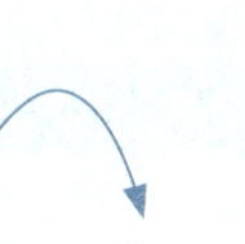
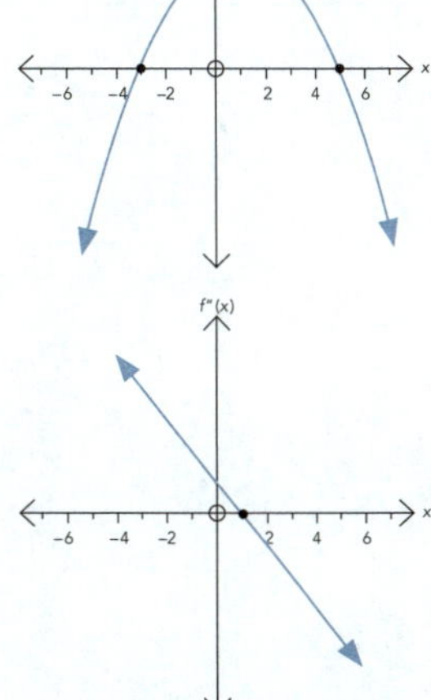
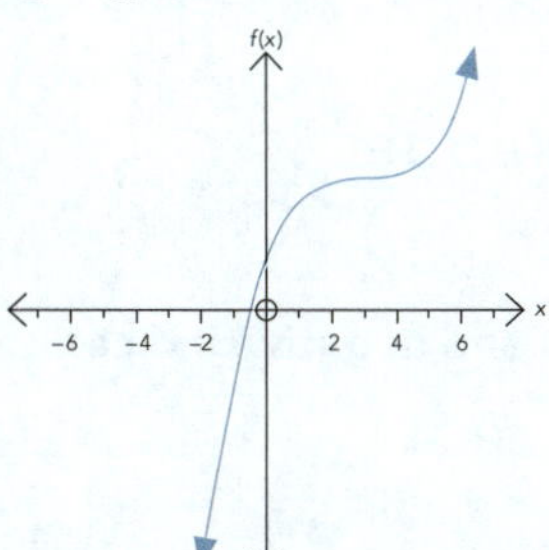

b $x = -3, 5$

c $-3 < x < 5$

d $x > 1$

3 a

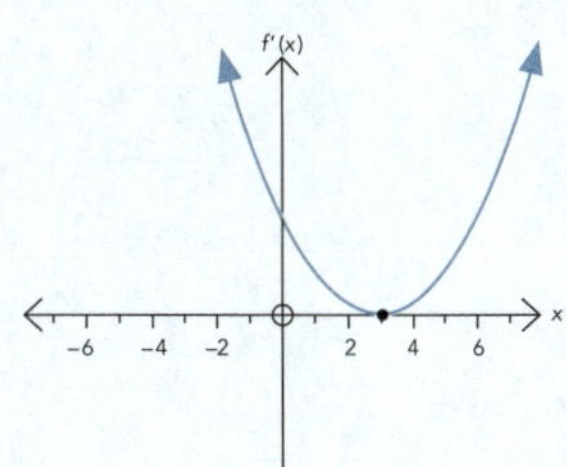
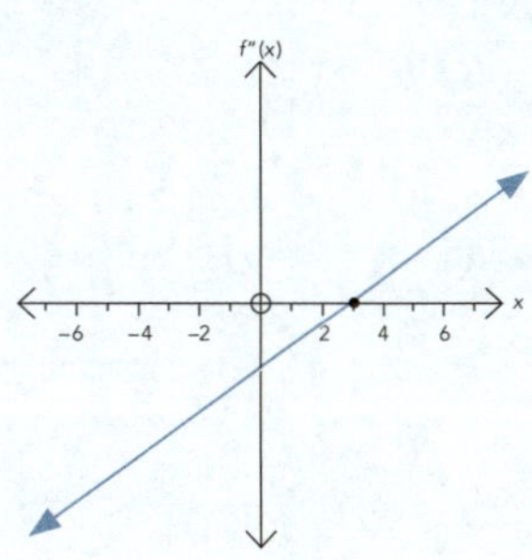

b $x = 3$

c No values

d $x \geq 3$

4 a

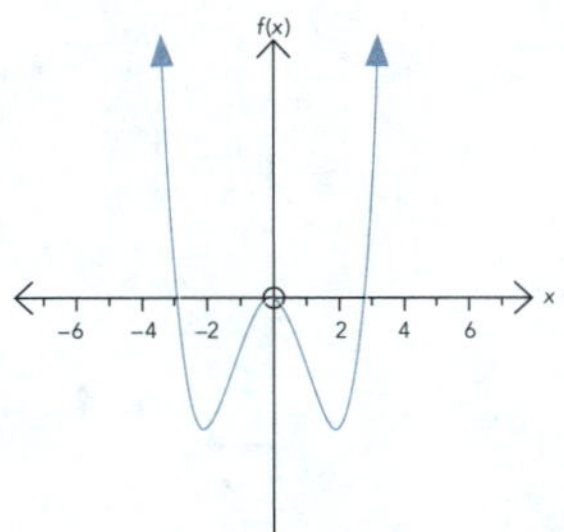
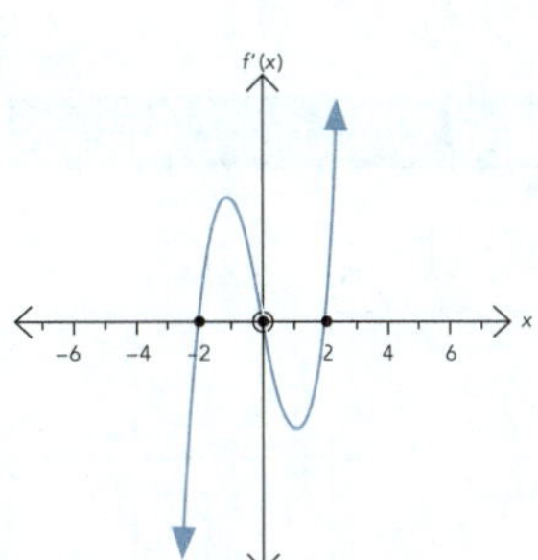
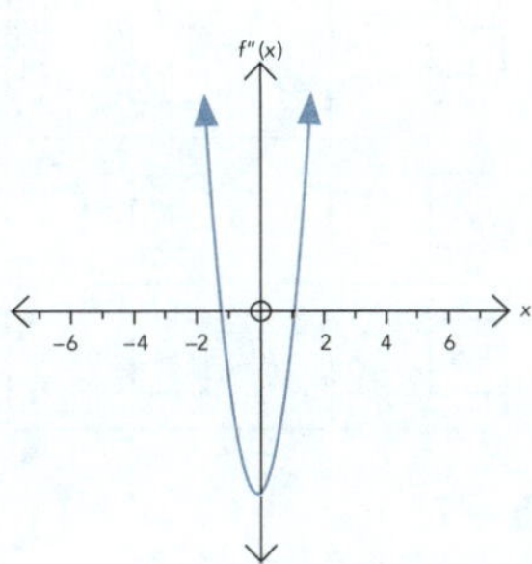

b $x = -2, 0, 2$

c $-2 < x < 0$ and $x > 2$

d $x < 0$

Piecewise functions (p. 20)

1 6
2 8
3 –3
4 0
5 2
6 4
7 4
8 undefined
9 undefined
10 5

Limits (pp. 21–22)

		Limit exists? (✓ or ×)	Value of limit if it exists	Value of $f(x)$ if it exists
1	$\lim_{x \to 2} f(x)$	✓	$f(x) = 5$	$f(x) = 1$
2	$\lim_{x \to 0} f(x)$	×	Does not exist	$f(x) = 3$
3	$\lim_{x \to 3} f(x)$	×	Does not exist	Does not exist
4	$\lim_{x \to -3} f(x)$	×	Does not exist	$f(x) = 2$

ISBN: 9780170446976

Continuous functions (p. 23)

1	$x = 1$	✗	**2**	$x = 0$	✓
	$x = 2$	✓		$x = 1$	✗
3	$x = 0°$	✓	**4**	$x = -2$	✓
	$x = 90°$	✗		$x = 1$	✓

Differentiability (p. 24)

	x value	Defined?	Continuous?	Differentiable?
1	$x = 1$	✓	✓	✗
	$x = 2$	✓	✗	✗
2	$x = 0$	✓	✓	✓
	$x = 3$	✗	✗	✗
3	$x = 0$	✓	✓	✓
	$x = -2$	✗	✗	✗

Putting it all together (pp. 25–27)

1

	Defined at $x = 2$?	$f(2)$	Limit exists?	$\lim\limits_{x \to 2}$	Continuous at $x = 2$?	Differentiable at $x = 2$?
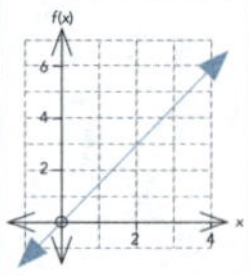	✓	3	Exists	3	✓	✓
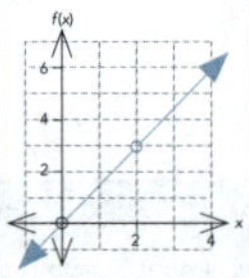	✗	Undefined	Exists	3	✗	✗
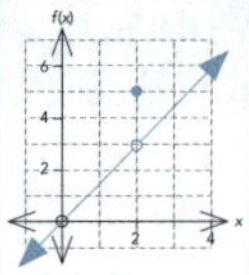	✓	5	Exists	3	✗	✗
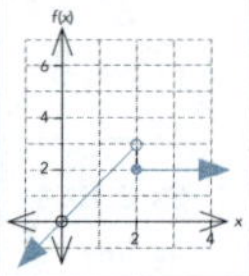	✓	2	Does not exist	Does not exist	✗	✗
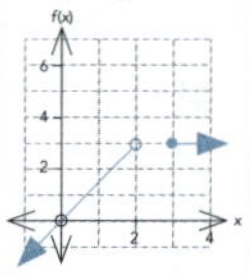	✗	Undefined	Does not exist	Does not exist	✗	✗
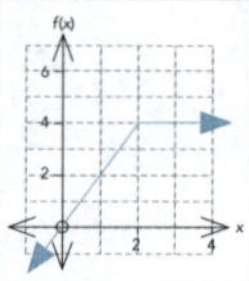	✓	4	Exists	4	✓	✗
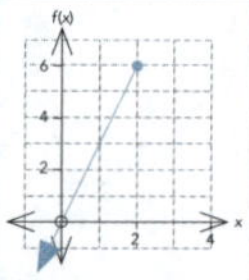	✓	6	Does not exist	Does not exist	✗	✗
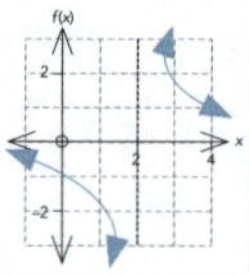	✗	Undefined	Does not exist	Does not exist	✗	✗

2 **a** **i** $x < -6$ and $x = -2$

ii $x = -3, 2$

iii $-2 < x < 0$

iv $x = -6, 0$

b 1 **c** $f(x) = 1$

3 **a** **i** $-3 \le x < -2,\ x = 1$

ii $x < -3,\ -2 \le x < -1,\ x > 4$

iii $x = -1$

b 3 **c** 1

d Does not exist

Differentiation techniques (pp. 28–50)

1 Expressions with fractional and negative indices (pp. 28–29)

1 $f'(x) = -4x^{-5}$ or $\dfrac{-4}{x^5}$

2 $\dfrac{dy}{dx} = \dfrac{1}{2} \,.\, 6x^{-\frac{1}{2}}$ or $\dfrac{dy}{dx} = \dfrac{3}{\sqrt{x}}$

3 $\dfrac{dy}{dx} = \dfrac{5}{3} \,.\, x^{-\frac{2}{3}}$ or $\dfrac{dy}{dx} = \dfrac{5}{3(\sqrt[3]{x^2})}$

4 $f'(x) = -\dfrac{1}{2}x^{-2}$ or $f'(x) = -\dfrac{1}{2x^2}$

5 $f'(x) = -10x^{-6}$ or $f'(x) = -\dfrac{10}{x^6}$

6 $\dfrac{dy}{dx} = -\dfrac{1}{2}x^{-\frac{3}{2}}$ or $\dfrac{dy}{dx} = -\dfrac{1}{2(\sqrt{x^3})}$

7 $\dfrac{dy}{dx} = -2 \,.\, \dfrac{3}{4} \,.\, x^{-3}$ or $\dfrac{dy}{dx} = -\dfrac{3}{2x^3}$

8 $f'(x) = 6 \,.\, \dfrac{2}{3} \,.\, x^{-\frac{1}{3}}$ or $f'(x) = \dfrac{4}{\sqrt[3]{x}}$

9 $f'(x) = \dfrac{3\sqrt{x}}{2}$

10 $\dfrac{dy}{dx} = \dfrac{2}{5} \,.\, \dfrac{5}{2} \,.\, x^{\frac{3}{2}}$ or $\dfrac{dy}{dx} = x\sqrt{x}$

ISBN: 9780170446976

11 $\frac{dy}{dx} = \frac{8}{3} \cdot -\frac{1}{4} \cdot x^{-\frac{5}{4}}$ or $\frac{dy}{dx} = -\frac{2}{3(\sqrt[4]{x^5})}$

12 $f'(x) = \frac{4}{3} \cdot -\frac{5}{2} \cdot x^{-\frac{7}{2}}$ or $f'(x) = -\frac{10}{3x^3\sqrt{x}}$

2 The chain rule (pp. 30–32)

1 $f'(x) = 2(x + 4)^1$ or $f'(x) = 2x + 8$

2 $\frac{dy}{dx} = 12(x^2 - 3)^2 \cdot 2x$ or $24x(x^2 - 3)^2$

3 $\frac{dy}{dx} = 10(8 - 3x^2) \cdot (6x)$ or $-60x(8 - 3x^2)$

4 $f'(x) = 24(x^2 - 2x)^3 \cdot (2x - 2)$ or $48(x^2 - 2x)^3(x - 1)$

5 $f'(x) = 4(2x^3 - x + 10)^3 \cdot (6x^2 - 1)$

6 $\frac{dy}{dx} = 7(x^2 + 3x + 4)^6 \cdot (2x + 3)$

7 $\frac{dy}{dx} = \frac{1}{2}(5x + 1)^{-\frac{1}{2}} \cdot 5$ or $\frac{dy}{dx} = \frac{5}{2\sqrt{(5x+1)}}$

8 $f'(x) = -1(2x - 1)^{-2} \cdot 2$ or $f'(x) = -\frac{2}{(2x-1)^2}$

9 $f'(x) = -2(5x + 1)^{-3} \cdot 5$ or $f'(x) = -\frac{10}{(5x+1)^3}$

10 $\frac{dy}{dx} = -8(x - 3)^{-3}$ or $\frac{dy}{dx} = -\frac{8}{(x-3)^3}$

11 $\frac{dy}{dx} = -3(3 - x)^{-\frac{3}{2}} \cdot -1$ or $\frac{dy}{dx} = \frac{3}{\sqrt{(3-x)^3}}$

12 $f'(x) = \frac{1}{2}(2x^4 - 5)^{-\frac{1}{2}} \cdot 8x^3$ or $\frac{4x^3}{\sqrt{2x^4 - 5}}$

3 Exponential functions (pp. 33–34)

1 $\frac{dy}{dx} = 5 \cdot e^{5x}$ or $\frac{dy}{dx} = 5e^{5x}$

2 $f'(x) = -1 \cdot e^{-x} = -\frac{1}{e^x}$

3 $f'(x) = 1 \cdot e^{x-3} = f'(x) = e^{x-3}$

4 $\frac{dy}{dx} = 2x \cdot e^{x^2} = 2xe^{x^2}$

5 $\frac{dy}{dx} = -7 \cdot e^{2-7x}$ or $\frac{dy}{dx} = -7e^{2-7x}$

6 $f'(x) = 3 \cdot 4e^{3x} = 12e^{3x}$

7 $f'(x) = 10 \cdot 2x \cdot 3^{x^2-3} = 20xe^{x^2-3}$

8 $\frac{dy}{dx} = 2 \cdot (6x + 1) \cdot e^{3x^2+x} = 2(6x + 1)e^{3x^2+x}$

9 $f'(x) = -2x^{-3} \cdot e^{\frac{1}{x^2}} = \frac{-2}{x^3} e^{\frac{1}{x^2}}$

10 $\frac{dy}{dx} = -2 \cdot e^{-2x} = \frac{-2}{e^{2x}}$

11 $\frac{dy}{dx} = 15 \cdot e^{5x} + 1 = 15e^{5x} + 1$

12 $f'(x) = 2 \cdot (2x - 8)\, e^{(x-4)^2} = 4(x - 4)e^{(x-4)^2}$

4 Logarithmic functions (pp. 35–36)

1 $\frac{dy}{dx} = 6 \cdot \frac{1}{6x}$
$= \frac{1}{x}$

2 $f'(x) = 7 \cdot \frac{1}{7x+2}$
$= \frac{7}{7x+2}$

3 $f'(x) = 5 \cdot \frac{1}{5x}$
$= \frac{1}{x}$

4 $\frac{dy}{dx} = 4x^3 \cdot \frac{1}{x^4}$
$= \frac{4}{x}$

5 $f'(x) = 15x^4 \cdot \frac{1}{3x^5}$
$= \frac{5}{x}$

6 $\frac{dy}{dx} = 2 \cdot 7 \cdot \frac{1}{7x}$
$= \frac{2}{x}$

7 $\frac{dy}{dx} = 4 \cdot 2x \cdot \frac{1}{x^2 - 1}$
$= \frac{8x}{x^2 - 1}$

8 $\frac{dy}{dx} = (6x^2 + 1) \cdot \frac{1}{2x^3 + x}$
$= \frac{6x^2 + 1}{2x^3 + x}$

9 $f'(x) = 14x + 2 \cdot \frac{1}{x}$
$= 14x + \frac{2}{x}$

10 $\frac{dy}{dx} = 5 \cdot 9 \cdot \frac{1}{9x - 2}$
$= \frac{45}{9x - 2}$

5 Trigonometric functions (pp. 37–39)

Trigonometric functions other than sine, cosine and tangent

1 $y = 2\sec(x)$

2 $f(x) = x\cot(3x)$

3 $f(x) = \tan(6x)$

4 $y = 9\cos(x)$

Differentiation of trigonometric functions

1 $\frac{dy}{dx} = 2\cos(2x)$

2 $f'(x) = 7\sin(7x)$

3 $\frac{dy}{dx} = -5\,\text{cosec}^2(5x)$

4 $f'(x) = 3\sec(3x)\tan(3x)$

5 $f'(x) = 8\sec^2(8x - 3)$

6 $f'(x) = -2x\,\text{cosec}(x^2)\cot(x^2)$

7 $\frac{dy}{dx} = -36x^2\,\text{cosec}^2(2x^3)$

8 $\frac{dy}{dx} = -2\sec(-2x)\tan(-2x)$

6 The product rule (pp. 40–44)

1 $f'(x) = x^5 . 4 + (4x - 5) . 5x^4$ or $f'(x) = 4x^5 + 5x^4(4x - 5)$

2 $f'(x) = (3x + 1) . 1 + (x - 7) . 3$ or
$f'(x) = (3x + 1) + 3(x - 7)$

3 $\frac{dy}{dx} = (8x + 5) . -2 + (3 - 2x) . 8$ or
$\frac{dy}{dx} = -2(8x + 5) + 8(3 - 2x)$

4 $\frac{dy}{dx} = (x^2 - 6) . 7 + (7x + 1) . 2x$ or
$\frac{dy}{dx} = 7(x^2 - 6) + 2x(7x + 1)$

5 $f'(x) = \sqrt{x} . 9 + (9x - 2) . \frac{1}{2}x^{-\frac{1}{2}}$ or
$f'(x) = 9\sqrt{x} + \frac{9x - 2}{2\sqrt{x}}$

6 $f'(x) = (3x - 5) . (2x + 4) + (x^2 + 4x - 1) . 3$ or
$f'(x) = (2x + 4)(3x - 5) + 3(x^2 + 4x - 1)$

7 $f'(x) = x^3 . e^x + e^x . 3x^2$ or $x^3e^x + 3x^2e^x$

8 $f'(x) = 5x^4 . 2e^{2x} + e^{2x} . 20x^3 = 10x^4e^{2x} + 20x^3e^{2x}$

9 $\frac{dy}{dx} = e^{7x} . 2x + (x^2 + 4) . 7e^{7x} = 2xe^{7x} + 7e^{7x}(x^2 + 4)$

10 $\frac{dy}{dx} = e^{2x} . \cos(x) + \sin(x) . 2e^{2x} = e^{2x}\cos(x) + 2e^{2x}(\sin x)$

11 $f'(x) = e^x . (8x - 3) + (4x^2 - 3x + 1) . e^x$
$= e^x(8x - 3) + e^x(4x^2 - 3x + 1)$

12 $f'(x) = e^{2x} . 3(5x - 9)^2 . 5 + (5x - 9)^3 . 2e^{2x}$
$= 15e^{2x}(5x - 9)^2 + 2e^{2x}(5x - 9)^3$

13 $f'(x) = e^{3x^2} . 2(2x - 1)^2 . 2 + (2x - 1)^2 . 6xe^{3x^2}$ or
$4e^{3x^2}(2x - 1) + 6x(2x - 1)^2e^{3x^2}$

14 $\frac{dy}{dx} = e^{\sqrt{x}} . (6x^2 - 5) + (2x^3 - 5x + 1) . \frac{1}{2\sqrt{x}} . e^{\sqrt{x}}$
$= e^{\sqrt{x}}(6x^2 - 5) + \frac{e^{\sqrt{x}}}{2(\sqrt{x})}(2x^3 - 5x + 1)$

15 $f'(x) = x . \frac{1}{x} + \ln(2x) . 1$ or $1 + \ln(2x)$

16 $\frac{dy}{dx} = \sin(x) . 6 . \frac{1}{6x} + \ln(6x) . \cos(x)$ or
$\frac{\sin(x)}{x} + \ln(6x)\cos(x)$

17 $\frac{dy}{dx} = x^3 . 2 . \frac{1}{2x - 5} + \ln(2x - 5) . 3x^2$
$= \frac{2x^3}{2x - 5} + 3x^2\ln(2x - 5)$

18 $f'(x) = 3 + 4x . 2x . \frac{1}{x^2} + \ln(x^2) . 4$
$= 11 + 4\ln(x^2)$

19 $f'(x) = 20 - 15x . 3 . \frac{1}{3x} - \ln(3x) . 15)$
$= 5 - 15\ln(3x)$

20 $\frac{dy}{dx} = e^x . 5 . \frac{1}{5x} + \ln(5x) . e^x$
$= \frac{e^x}{x} + e^x\ln(5x)$

21 $f'(x) = x^2 . \cos(x) + \sin(x) . 2x$
$= 2x\sin(x) + x^2\cos(x)$

22 $\frac{dy}{dx} = 2x^3 . 3\text{cosec}(3x)\cot(3x) + \text{cosec}(3x) . 6x^2$
$= -6x^3\text{cosec}(3x)\cot(3x) + 6x^2\text{cosec}(3x)$

23 $f'(x) = e^x . 5 . \sec(5x)\tan(5x) + \sec(5x) . e^x$
$= 5e^x\sec(5x)\tan(5x) + e^x\sec(5x)$

24 $f'(x) = (7x + 2) . 3 . -\sin(3x) + \cos(3x) . 7$
$= 7\cos(3x) - 3(7x + 2)\sin(3x)$

25 $\frac{dy}{dx} = e^{4x} . 5\sec^2(5x) + \tan(5x) . 4e^{4x}$ or
$5e^{4x}\sec^2(5x) + 4e^{4x}\tan(5x)$

26 $f'(x) = (3x - 1)^2 . -2\sin(2x) + \cos(2x) . 2(3x - 1) . 3$ or
$6(3x - 1)\cos(2x) - 2(3x - 1)^2\sin(2x)$

27 $\frac{dy}{dx} = (2x - 3)^5 . 2(x + 4) . 1 + (x + 4)^2 . 5(2x - 3)^4 . 2$
or $2(2x - 3)^5(x + 4) + 10(x + 4)^2(2x - 3)^4$

28 $\frac{dy}{dx} = (5x - 2)^3 . 2x + (x^2 + 1) . 3(5x - 2)^2 . 5$
or $2x(5x - 2)^3 + 15(x^2 + 1)(5x - 2)^2$

29 $\frac{dy}{dx} = (x + 2)^3 . -\sin(x) + \cos(x) . 3(x + 2)^2$ or
$3(x + 2)^2\cos(x) - (x + 2)^3\sin(x)$

30 $\frac{dy}{dx} = 6x . \frac{1}{2}x^{-\frac{1}{2}} . e^{\sqrt{x}} + e^{\sqrt{x}} . 6$ or
$\frac{3xe^{\sqrt{x}}}{\sqrt{x}} + 6e^{\sqrt{x}}$

7 The quotient rule (pp. 45–47)

1 $\frac{dy}{dx} = \frac{(x - 3) . 1 - (x + 5) . 1}{(x - 3)^2} = \frac{-8}{(x - 3)^2}$

2 $f'(x) = \frac{(x + 4) . 2 - (2x - 7) . 1}{(x + 4)^2} = \frac{15}{(x + 4)^2}$

3 $\frac{dy}{dx} = \frac{(x - 1) . 8 - 8x . 1}{(x - 1)^2}$ or $\frac{dy}{dx} = \frac{-8}{(x - 1)^2}$

4 $f'(x) = \frac{(5x + 2) . 2x - x^2 . 5}{(5x + 2)^2} = \frac{5x^2 + 4x}{(5x + 2)^2}$

5 $\frac{dy}{dx} = \frac{\sin(x) . 1 - x . \cos(x)}{\sin^2(x)}$
or $\frac{\sin(x) - x\cos(x)}{\sin^2(x)}$

6 $\frac{dy}{dx} = \frac{\cos(x) . 2 - 2x . -\sin(x)}{\cos^2(x)}$
$= \frac{2\cos(x) + 2x\sin(x)}{\cos^2(x)}$

7 $\frac{dy}{dx} = \frac{\sin(x) . e^x - e^x\cos(x)}{\sin^2(x)}$
$= \frac{e^x\sin(x) - x\cos(x)}{\sin^2(x)}$

8 $\frac{dy}{dx} = \frac{x^2 . 2\sec^2(x) - 2\tan(x) . 2x}{x^4}$
or $\frac{2x^2\sec^2(x) - 4x\tan(x)}{x^4}$

9 $f'(x) = \frac{2x . 2(x - 3) . 1 - (x - 3)^2 . 2}{4x^2}$
or $\frac{x^2 - 9}{2x^2}$

ISBN: 9780170446976

10 $\frac{dy}{dx} = \frac{(2x-1)^2 . 8 - 8x . 2(2x-1) . 2}{(2x-4)^4}$

or $\frac{-8(2x+1)}{(2x-1)^3}$

11 $\frac{dy}{dx} = \frac{2x . e^x - e^x . 2}{4x^2}$ or $\frac{xe^x - e^x}{2x^2}$

12 $f'(x) = \frac{(x-3) . 5e^{5x} - e^{5x} . 1}{(x-3)^2}$ or $\frac{e^{5x}(5x-16)}{(x-3)^2}$

13 $\frac{dy}{dx} = \frac{e^{3x} . 1 - x . 3e^{3x}}{e^{6x}}$ or $\frac{1-3x}{e^{3x}}$

14 $\frac{dy}{dx} = \frac{(4-7x) . 2xe^{x^2} - e^{x^2} . -7}{(4-7x)^2}$

or $\frac{e^{x^2}(8x - 14x^2 + 7)}{(4-7x)^2}$

15 $f'(x) = \frac{\tan(x) . 6x - 3x^2 . \sec^2(x)}{\tan^2(x)}$

or $\frac{6x\tan(x) - 3x^2\sec^2(x)}{\tan^2(x)}$

16 $\frac{dy}{dx} = \frac{x^3 . -2\sin(2x) - \cos(2x) . 3x^2}{x^6}$

or $\frac{-2x\sin(2x) - 3\cos(2x)}{x^4}$

17 $\frac{dy}{dx} = \frac{\tan(3x) . 2e^{2x} - e^{2x} . 3\sec^2(3x)}{\tan^2(3x)}$

or $\frac{e^{2x} . (2\tan(3x) - 3\sec^2(3x))}{\tan^2(3x)}$

18 $f'(x) = \frac{e^{3x-2} . 5\cos(5x) - \sin(5x) . 3e^{3x-2}}{e^{6x-4}}$

or $\frac{e^{3x-2}(5\cos(5x) - 3\sin(5x))}{e^{6x-4}}$

8 Parametric functions (pp. 48–50)

1 $\frac{dy}{dx} = \frac{6t}{5}$ $\frac{d^2y}{dx^2} = \frac{6t}{25}$

2 $\frac{dy}{dx} = 6t$ $\frac{d^2y}{dx^2} = \frac{3}{t}$

3 $\frac{dy}{dx} = 12t^3$ $\frac{d^2y}{dx^2} = 36t^3$

4 $\frac{dy}{dx} = -5t^4$ $\frac{d^2y}{dx^2} = \frac{5t^7}{2}$

5 $\frac{dy}{dx} = \frac{6t^3}{e^{2t}}$

6 $\frac{dy}{dx} = \frac{1}{4t\sqrt{t+3}}$

7 $\frac{dy}{dx} = \frac{e^{2t}}{t^2}$

8 $\frac{dy}{dx} = 6t^3 - 2t^2$

9 $\frac{dy}{dx} = \frac{-2\sin(2t)}{3\cos(t)}$

10 $\frac{dy}{dx} = \frac{2}{3}\cos^3(t)$

11 $\frac{dy}{dx} = -\frac{2}{\cos(t)}$

12 $\frac{dy}{dx} = \frac{2\sec(t)}{\tan(t)}$

Applications (pp. 51–89)

Tangents and normals (pp. 51–60)

1 Finding gradients and equations of tangents and normals (pp. 51–56)

1 $m = -12$
2 $y = 1$
3 $y = \frac{-x}{5} - \frac{14}{5}$
4 $m = -4$
5 m is undefined
6 $y = 3x - 1$
7 $m = -\frac{1}{4}$
8 $y = 1.693x - 1$
9 $m = 1$
10 $m = -1$
11 $m = 2.571$
12 $m = 18$
13 $m = -2$
14 $m = 2$
15 $m = 2$
16 m is undefined

2 Finding coordinates of points given the equation and gradient (pp. 57–60)

1 $x = 0.1733$
2 $x = \frac{1}{2}$
3 (21, −3) and (21, 3)
4 $x = -2$ and 6
5 (−2, 0)
6 $\left(\frac{\pi}{3}, 0.5\right)$
7 (3, 3) and (7, 19)
8 $x = -7$
9 $x = 3$

Stationary points (pp. 61–72)

1 Finding stationary points (pp. 61–64)

1 **a** Stationary points at (0, 1) and (1, 0).
b $f''(0) = 0 \Rightarrow$ point of inflection at (0, 1)
$f''(1) = \frac{2}{3} \Rightarrow$ minimum at (1, 0)
Because there is a minimum at (1, 0), the point of inflection at (0, 1) must be decreasing.

2 **a** $x = 2.95$
b $f''(2.95) = -44.4 \Rightarrow$ maximum point
Coordinates: (2.95, −66.19)

3 **a** $k = 3.718$
b $\frac{d^2y}{dx^2} = 10.15 \Rightarrow$ minimum point
Coordinates: (1, 7.437)

4 $x = -2$ and 8
5 $x = 2$

2 Optimisation (pp. 65–72)

1 Minimum on day 5, when 20 insects remained
2 Maximum area = 16 units^2, vertex at (2, 16)
3 Turning point at (12, 28). Maximum. The maximum rate of growth of the tree occurs when it is 12 years old, and after that its rate of growth declines.
4 (4, 4)
5 Maximum area = 16 square units, A(−2, 8), B(2, 8)
6 Average number of staff = 4.482
7 $r = x = 28$ m
8 Maximum area = 72 square units, B($3\sqrt{2}$, $3\sqrt{2}$)
9 $r = 1.128$ m
10 Maximum volume = $16\sqrt{3} = 20.78$ cm^3, $x = 4$ cm and length = 4 cm

ISBN: 9780170446976

Related rates (pp. 73–85)

1 Questions involving two rates (pp. 73–76)

A

1 $\frac{dV}{dr} = \frac{dt}{dr} \times \frac{dV}{dt}$

2 $\frac{dr}{dt} = \frac{dV}{dt} \times \frac{dr}{dV}$

3 $\frac{dA}{dt} = \frac{dx}{dt} \times \frac{dA}{dx}$

4 $\frac{dP}{dt} = \frac{dx}{dt} \times \frac{dP}{dx}$

5 $\frac{dV}{dx} = \frac{dV}{dr} \times \frac{dr}{dx}$

6 $\frac{dP}{dt} = \frac{dx}{dt} \times \frac{dP}{dx}$

7 $\frac{dA}{dt} = \frac{dx}{dt} \times \frac{dA}{dx}$

8 $\frac{dS}{da} = \frac{dS}{dx} \times \frac{dx}{da}$

9 $\frac{dV}{da} = \frac{dt}{da} \times \frac{dV}{dt}$

10 $\frac{dA}{dr} = \frac{dx}{dr} \times \frac{dA}{dx}$

11 $\frac{dV}{dr} = \frac{dt}{dx} \times \frac{dr}{dt} = \frac{dV}{dx}$

12 $\frac{dx}{dr} \times \frac{dA}{dx} \times \frac{dr}{dt} = \frac{dA}{dt}$

13 $\frac{dA}{dt} = \frac{dV}{dt} \times \frac{dr}{dV} \times \frac{dA}{dr}$

14 $\frac{dS}{dr} = \frac{dt}{dr} \times \frac{dx}{dt} \times \frac{dS}{dx}$

15 $\frac{dV}{dt} = \frac{dr}{dx} \times \frac{dV}{dr} \times \frac{dx}{dt}$

16 $\frac{dA}{dr} = \frac{dA}{ds} \times \frac{dt}{dr} \times \frac{ds}{dt}$

17 $\frac{dP}{dr} = \frac{dt}{dx} \times \frac{dx}{dr} \times \frac{dP}{dt}$

18 $\frac{dS}{da} = \frac{dr}{dt} \times \frac{dS}{dr} \times \frac{dt}{da}$

B Questions involving two rates (pp. 74–76)

1 120 mm/s

2 –10.8 mm³/min

3 480π cm²/s or 1508 cm²/s (4 sf)

4 $\frac{2}{25\pi}$ or 0.1528 cm/s

5 70.56 cm³/min

6 $\frac{5}{144\pi} = 0.1105$ cm/s

2 Questions involving ratios (pp. 77–81)

1 –0.016 radians/s

2 3.2 m/s

3 0.48 m/min

4 $0.2\dot{6}$ m/s

5 0.0600 radian/s

6 0.48 m/s

3 Questions involving three rates (pp. 82–85)

1 200 cm²/s

2 13.25 cm³/s

3 4.675 mm³/h

4 2.771 m²/min

5 5.625 mm³/week

Kinematics (pp. 86–89)

1 $t = 1.\dot{3}$ s and $t = 3$ s

2 $t = \sqrt{2}$ s

3 $v(t) = 0.\dot{8}\dot{6}\dot{3}$ m/s

4 $a(t) = \frac{1 - 0.5x}{x - 0.25x}$

$a(2) = \frac{1-1}{2-1} = 0$ m/s²

5 $t = \frac{1}{2}\ln\frac{5}{2} = 0.4581$

Practice questions (pp. 90–95)

Practice question 1 (pp. 90–91)

a $\frac{dy}{dx} = -4\text{cosec}^2\theta$

b $\frac{dy}{dx} = \frac{2}{t^3\sin t}$

c $\frac{dy}{dx} = -2\ln x - 2x \cdot \frac{1}{x}$

$= -2\ln x - 2$

Turning point $\Rightarrow \frac{dy}{dx} = 0$

$\therefore \ln x = -1$

$x = e^{-1} = 0.368$, so $y = 5.736$

$\frac{d^2y}{dx^2} = -\frac{2}{x} = -\frac{2}{0.368} \Rightarrow$ Maximum

d $\frac{d\theta}{dt} = \frac{dh}{dt} \cdot \frac{d\theta}{dh}$

$= 1.5 \cdot \frac{\cos^2\theta}{20}$

$= 1.5 \cdot \frac{(\cos 0.4636)^2}{20}$

$= 0.06$ radians

$\tan\theta = \frac{h}{20} \Rightarrow h = 20\tan\theta$

$\frac{dh}{d\theta} = 20\sec^2\theta$

$= \frac{20}{\cos^2\theta}$

$h = 10 \Rightarrow \theta = 0.4636$

e $\frac{dy}{dx} = e^{-(x-2)^2} \cdot (-2x + 4)$

$= -e^{-(x-2)^2} \cdot (2x - 4)$

$\frac{d^2x}{dx^2} = [-e^{-(x-2)^2} \cdot 2] + [(2x - 4) \cdot e^{-(x-2)^2} \cdot (4 - 2x)]$

$= -e^{-(x-2)^2}(4x^2 - 16x + 14)$

$= -2e^{-(x-2)^2}(2x^2 - 8x + 7)$

$-2e^{-(x-2)^2}$ cannot equal 0

$\therefore (2x^2 - 8x + 7) = 0$

$\Rightarrow x = 1.2929$ or 2.7071

Practice question 2 (pp. 92–93)

a $\frac{dy}{dx} = \frac{-6}{x^4} + \frac{1}{2\sqrt{x^3}}$

b $f'(t) = 1.5$

c **i** $y = 6$

ii $x = -4$

iii **1** $x = 4$

2 $x < -4, -2 < x < 0, x > 4$

iv 6

d $a = \frac{dv}{dt}$

$= 3 + (5e^{-2t} \cdot -2) = 0$

$\therefore 10e^{-2t} = 3$

$e^{-2t} = 0.3$

$t = -\frac{\ln 0.30}{2} = 0.602$ seconds

 ISBN: 9780170446976

e $V = \frac{1}{3}\pi r^2 h$

$= \frac{1}{3}\pi r^2 (6 - r)$

$= 2\pi r^2 - \frac{1}{3}\pi r^3$

$\frac{dV}{dr} = 4\pi r - \pi r^2$

$\Rightarrow r(4 - r) = 0$

Maximum $\Rightarrow r = 4$, so $V = 33.51$ cm^3

Practice question 3 (pp. 94–95)

a $f'(x) = 6xe^{5x} + 5e^{5x}(3x^2 - 7)$

b $m = \frac{\sqrt{x^3}}{4}$

c Area $= xy = x(6 - \sqrt{x})$

$= 6x - x^{\frac{3}{2}}$

Maximum where $\frac{dA}{dx} = 6 - \frac{3}{2}\sqrt{x} = 0$

$\sqrt{x} = 4$

$x = 16$

$\therefore y = 6 - \sqrt{16} = 2$

Maximum area $= 16(6 - 4) = 32$ at $(16, 2)$

d $\frac{dy}{dx} = \frac{\cos x \,.\, 2e^{2x} + e^{2x} \,.\, \sin x}{\cos^2 x}$

$= \frac{e^{2x}}{\cos x}\left(\frac{2\cos x + \sin x}{\cos x}\right)$

$= y(2 + \tan x)$

e $\frac{dA}{dt} = \frac{dV}{dt} \cdot \frac{dr}{dV} \cdot \frac{dA}{dr}$

$= \frac{12}{1} \cdot \frac{1}{4\pi r^2} \cdot \frac{8\pi r}{1}$

$= \frac{24}{r}$

$r = 9 \Rightarrow \frac{dA}{dt} = 2.\dot{6}$ mm^2/s

$V = \frac{4}{3}\pi r^3$

$\frac{dV}{dr} = 4\pi r^2$

$A = 4\pi r^2$

$\frac{dA}{dr} = 8\pi r$